Thirty Years with Stata:
A Retrospective

Thirty Years with Stata: A Retrospective

Edited by
ENRIQUE PINZON
StataCorp

A Stata Press Publication
StataCorp LP
College Station, Texas

® Copyright © 2015 by StataCorp LP
All rights reserved. First edition 2015

Published by Stata Press, 4905 Lakeway Drive, College Station, Texas 77845
Typeset in LaTeX 2_ε
Printed in the United States of America

10 9 8 7 6 5 4 3 2 1

ISBN-10: 1-59718-172-2
ISBN-13: 978-1-59718-172-3

Library of Congress Control Number: 2015940894

Contents

Contents ix

Preface

Thirty years ago, Stata was created with a vision of a future in which computational resources would make data management and statistical analysis readily available and understandable to someone with a personal computer. The research landscape today validates this vision. We have increasing amounts of data and computational power and are faced with the necessity for rigorous statistical analysis and careful handling of our information.

Yet the increasing amount of information also poses daunting challenges. We have to disentangle the part of our data that communicates meaningful relationships (signal) from the part that is uninformative (noise). Distinguishing signal from noise was the task of researchers thirty years ago, and it is still today. The fact that we have more data and computational prowess does not mean that we have less noise lurking. The tools of careful data management, programming, and statistical analysis are fundamental to maximize the potential of our resources and to minimize the noise.

In this volume, we gather 14 essays and an interview that answer how Stata has helped researchers in the process of distinguishing signal from noise. In the first part, we begin with a new essay based on a speech given by Bill Gould during Stata's 2014 holiday celebration, which we follow by revisiting two contributions that were written to commemorate Stata's 20th anniversary. These three pieces provide a perspective from inside Stata. The first, the speech, discusses the decisions made regarding Stata's software architecture. The second is an interview of Bill Gould by Joe Newton that reveals the guiding principles behind Stata. The third, written by Sean Becketti, gives us insight into the culture and challenges of Stata in its developing stages.

The second part of the book represents points of view from the outside. Researchers from different disciplines answer how Stata has helped advance research in their fields and how their fields have evolved in the past three decades. Some of the contributions look at the discipline as a whole, while others speak about very specific experiences with Stata. The contributions in this part come from the disciplines of behavioral science, business, economics, epidemiology, time series, political science, public health, public policy, veterinary epidemiology, and statistics. Also in this part, Nick Cox writes about the history of Stata and devotes part of his essay to the conception and evolution of the Stata User Group meetings.

Having a vision from inside and a vision from outside is a fitting way to celebrate Stata's 30th anniversary. The vision from inside reminds us of the ideas that made Stata popular and the principles that guide Stata to this day. Yet, it is the researchers, their interests, their concerns, their active participation, and their interaction with Stata that help the software evolve. The relationship between Stata users and StataCorp is the fundamental reason that we are celebrating this anniversary.

To all that have been inside and to all those outside, this book is for you.

Acknowledgments

The authors in this volume contributed to the final product not only with their chapters but also with their suggestions and input during the entire process.

Shelbi Seiner and Stephanie White did a wonderful job, respectively, with the editing and the formatting of the book. The cover design is the work of Annette Fett.

To all of you, thanks.

Caveat Emptor

All the contributions in this volume were written before the release of Stata 14.

Part I

A vision from inside

1 Initial thoughts

William Gould

It is 2015 and Stata is celebrating its 30th anniversary. Stata 1.0 was released in January of 1985. That is a long time ago. That Stata still survives is remarkable. That Stata is not a dinosaur is even more remarkable.

Stata was born in 1985. Some of you were not yet born. Others may not remember 1985 as clearly as they would wish. So let me set the stage.

Ronald Reagan was president. Mikhail Gorbachev became head of the Soviet Union. Ronald Reagan would not say "Tear down this wall" for another two and one-half years (Stata 1.5), and the wall would not fall for another two and one-half years after that (Stata 2.05). NASA discovered a hole in the ozone over the Antarctic. Boom boxes were popular. Among the top movies of 1985 was *Back to the Future*.

In 1985, the expensive desktop computers had five-and-a-quarter inch floppy drives. Cheaper ones had cassette tape. Memory was measured in kilobytes, and in only two digits at that.

In 1985, the software available for the desktops was crude. PC-DOS 3.1 was released. So was Windows 1.0, but it was more a forward-looking experimental demonstration system than anything else. It was not much used, and Windows 3.1 was still seven years away.

The first Mac was released in 1984, the year before the release of Stata 1.0. Work on Stata 1.0 began in 1984.

Stata was written in C, and that was an odd choice. The popular mainframe computer languages were PL/I, FORTRAN, COBOL, and various assemblers. The popular small-computer languages were assembler, BASIC, and PASCAL. All the buzz was about PASCAL. In 1984, Apple released PASCAL and Borland released TurboPASCAL.

1984 was the year of PASCAL and Stata was written in the wrong language, everybody told me. Real software, of course, was implemented in FORTRAN, but FORTRAN was not available for small computers, which just went to show that the small computers were not real computers. "Little computers for little minds" was a popular saying among real programmers.

Some real programmers worked on minicomputers, but even that was not fully respectable. A lot of that work involved something called ARPANET, which many believed to be going nowhere, and in 1985, that view was proven correct when ARPANET was transferred to the academic backwater of NSFNET. Later, NSFNET would develop into a major part of the Internet backbone.

Despite all of this, interest in small computers was growing, and the dead-enders from ARPANET escaped to Santa Barbara, Palo Alto, and Berkeley. And it was exciting. Nonexperts were coming out of the woodwork to write software. Stata was not the first statistical system for microcomputers, nor the second.

Stata did not have bright prospects, and it would not have had bright prospects even if it had been written in the right language. Because Stata survives, some might conclude that I am brilliant. Yet I did not predict the powerful computers that we have today, the demise of the mainframe, or the rise of the Internet. I merely thought that PCs were powerful enough to perform data analysis on smaller datasets and that they could do it cheaper and better.

I do not know what my partner, Finis Welch, was thinking when he trusted me enough to run with it. At that time, I did think that C was the language of the future, and I had strong opinions about Stata's design, opinions that would be reconstructed and refined with the collaboration of Sean Becketti.

I cannot take all the credit for Stata, but I am proud of two of my contributions: I was right about C, it was the language of the future. And I was right about Stata's design, which has proven flexible enough to accommodate all the other changes I did not predict.

2 A conversation with William Gould

H. Joseph Newton
Texas A&M University
jnewton@stat.tamu.edu

2.1 Beginnings of Stata

Newton: How did the first version of Stata come about, and what sorts of things could it do?

Gould: The first version of Stata was a regression package and really nothing more than that. It did a little bit in the way of calculations, and it did some summary statistics, but it was all built around a regression engine. It was written over a one-year period by me initially and by Sean Becketti, who helped me later. I wrote the C code; Sean Becketti helped me a lot with the design. I would say that half of the design is mine and half the design is Sean's in terms of what the user actually saw. A number of things became available just at that time when we started this project, and it was those things that actually caused the project to start. The first C compiler was available for the PC.

Newton: When was this?

Gould: 1984. This is our 20-year anniversary. Our official release date was 1985. The actual release date was December 1984. There was the *American Economic Association* meeting in Dallas. That was where we first announced Stata and started selling it. The meeting is usually around Christmas or New Year. I started about a year before that, so that puts the beginning date at about January of 1984. The Lattice C compiler had just become available for the PC, but no one cared because nobody was interested in C—everyone talked about Pascal as *the* language. I had learned C earlier, and I was very interested in C. No part of Stata was developed under Unix, but I was subtly familiar with the stuff, and so once those development tools became available, that really made it possible for me to write something like Stata.

Newton: When you started, did you have syntax, grammar, or anything like that in mind?

Gould: Yes, right from the beginning. For two reasons: The major reason—What was the term that [Brian W.] Kernighan used when describing the creation of Unix?—"Salvation through scarcity". Stata benefited exactly from that. These were really small

5

computers, 64K memory. You could not afford to have any replicated code. Everything had to happen just once so that the centralized parser and a centralized grammar of Stata could understand statements in the absence of what the particular command was or anything else. That was a really big feature in terms of the code, but that was also something that was right from the beginning a very important design criterion (something I really wanted to see in a package). I was a SAS user, and I had nothing but respect for SAS, and I still have nothing but respect for SAS, but in those days— and I suspect it is still true—anybody who was a real SAS programmer walked around carrying the SAS manual under their arm because nobody could remember how you were to do something. It varied according to the problem as to how you were supposed to do it.

Newton: Did your grammar come from reading other stuff, or did you make it up?

Gould: We made it up, and it was largely based on Wylbur as the basic grammar. Wylbur had a very simple grammar.

Newton: Wylbur, the old operating system?

Gould: Yes, it had a real simple editing type of grammar. And so you could say `list /1` and `list 5/10`. You will notice Stata has a `5/10` and `/1`. The keyword `in` was stuck in front of it because, unlike an editor, which had just one line, we had multiple columns per line per observation. And, so the grammar was made up just kind of willy-nilly, and then it went through improvements over time. There were a lot of unnecessary words in the earliest version of the grammar that only Sean Becketti ever saw. It was `replace x with 4`. There was a `with` keyword; there were a number of keywords that all got taken out and merged into a single equal sign. We overloaded the equal sign too far as we were cleaning up the language because we finally decided that weights could be done with just an equal sign. So you would say `regress miles-per-gallon length` or `regress something something` I don't care `= population`, and you just put an equal sign there because the expression was good enough to designate the weight. Of course, at that time I thought there was only one kind of weight, having never really thought about it, and a lot of people hadn't in those days. We borrowed from Wylbur; we borrowed from Unix, and we borrowed from CMS, the operating system at IBM, that is, IBM/CMS. There were some good features of all these, and we borrowed syntax left and right. Like I said, it was very wordy, and Sean and I working together got most of the words out of it.

Newton: So Becketti was a major force in the beginning stages?

Gould: When this thing started up, Finis Welch suggested that we bring Sean Becketti in; I think he basically just wanted us to bring somebody in. Finis Welch is my partner and was my advisor in graduate school. I learned all my economics, econometrics, and business from Finis; probably the most important person in my intellectual life.

2.2   Learning computing skills

Newton: Where did you learn your computing skills?

Gould: That had come earlier; I was into computers from the earliest age. In 8th grade was the first time I saw a computer for real. It was a really strange situation, SDC (System Development Corporation), which was a spinoff of Rand Corporation (I grew up in Santa Monica). They had a computer, and Q-32 was the number of it, and it had been designed to be used on the DEW line, the line of radars north of the US that were to be an early warning of Soviet missile attacks. I don't know if it ever was or whether that became dated before it ever happened, but this was a huge computer—one of the biggest in the world at the time with rooms upon rooms of vacuum tubes, and they developed an online system for it, and I guess they were trying to figure out some reason for this computer to exist. They got a government grant to be involved in teaching. They put two teletype terminals in our junior high school in Santa Monica one year. I learned how to program. I just thought it was the neatest thing.

Newton: What language?

Gould: TINT, and there was a pint-sized version of it called PINT. I can't remember what TINT and PINT stood for ... teletype ... interactive ....

Newton: Was it an operating system or basic programming?

Gould: At this time, there was not a big distinction between operating systems and applications. It was a big, monolithic thing. And, TINT was it; it was the language you talked to this online system. It was kind of a basic FORTRAN-type language.

Newton: Was it compiled?

Gould: No, it was an interpretative online system.

Newton: Have you ever used APL?

Gould: No.

Gould: Once I saw this computer and got to play with it, that summer during summer school I took typing. I think I was the only male in the class. Back then, it was women who took typing, and men who took other things. They took shop or something. And, there I was in this typing class learning how to type because I wanted to type on the teletype faster. From then on, it was amazing how, if you looked around you could find computers. I found one at my father's office, and then I learned FORTRAN. FORTRAN II was the first serious language I learned. When I got to high school, they had just gotten a 1620 computer in, and there was some guy they had hired who was trying to make it work, but he hadn't succeeded yet. I found this computer room with this computer in it, and a FORTRAN compiler, and so I made a reputation for myself as writing the first FORTRAN program that worked on this computer, and what it did wasn't much.

Newton: "Hello, world", or something?

Gould: Yeah, basically, or the equivalent. So I learned 1620 machine language in high school and FORTRAN better. I got my first real job in computers when I was a senior in high school. First as a, well ... there is a story that goes with it. In my junior year of high school, there was a dance coming up, and two friends and I decided we would write a computer dating program that would match up people. They would fill out a questionnaire, and so the committee went for it. They gave us a couple hundred dollars to do this. We wrote two versions of the program. One we wrote in FORTRAN II, and the other written in PL/I. We had disguised ourselves as college students and had gone down and used the Bolter Hall engineering computer at UCLA, and debugged our program there. It was kind of cool. People filled out their cards, and we typed it all in. The dance was the next night. We started our 1620 running the FORTRAN II program. It sat, and it blinked lights; and it sat, and it blinked lights. We were getting really kind of nervous. I don't know if you remember the IF SENSESWITCH. It was a statement of FORTRAN. So you throw this switch up, and as soon as the switch was thrown up, it started putting up debug information, and you could see where the program was. We figured out it was going to be finished sometime next week. We were in trouble. It was late at night that Thursday, and because SDC had done this thing with the junior high school, I said, "Let's go down there, and see if we can buy computer time from them." So, we go down there, and of course there are these guards at the desk. This is an aerospace firm. They are looking at these kids saying what are you guys doing here?

Newton: Sounds like a movie.

Gould: Yes, but the computer operator who was running the computer room that night said, "Sure, I'll run your stuff." He ran it! And, so we were leaving, and he was going to put it on at some point, and we didn't know how long it was going to take. He came running out to the parking lot just as we are getting in the car because it had run. This was like a 360–90 something; it got it done in a blink of an eye. Later, there was a little start-up firm that I had heard about that needed a computer operator, so I went down to apply there. And, who do I find there, but this same computer guy who is now working for this firm. He is Vice President of this little company. It's a spinoff. They are all guys from Rand or SDC. So I got a job there, and I worked as a computer operator and then as a programmer.

Newton: So you learned a lot of your computing on the job?

Gould: On the job. Absolutely! At that time, there were not computer science departments, there were a few computer science courses in the math department.

Newton: So you got Stata started, you and Finis Welch, as partners. Were you the only two partners?

Gould: We were the only two partners.

Newton: And Sean Becketti was involved?

Gould: Sean Becketti was hired to assist me, but he turned out to be very important. He was really great.

Newton: How long was he involved with Stata?

Gould: He worked for us for about two years while in LA, and then took on consulting status after that. We were a startup; I wanted him to stay, but he decided to continue his academic career. He moved to the Midwest and continued to be a consultant with us.

Newton: I know later he was one of the editors of the *Stata Technical Bulletin*. In fact, in the testimonials that accompany this interview, he is one of the people who had some things to say.

Gould: He sort of understated his contribution.

2.3   The CRC

Newton: So, you got it started, you're a couple of years down the road. What were you worried about then?

Gould: Sales! What you worry about is sales. You have to realize in those early days selling a copy was an exciting event for that whole day, selling two copies was unbelievable. Fortunately, the company that developed Stata, CRC, had a small mainframe operation, and that's where the money came from, so there wasn't a huge concern with Stata supporting itself, and there was no way that it was supporting itself.

Newton: Did CRC stand for something?

Gould: Computing Resource Center, and this goes back to the old time-sharing days.

Newton: Did it sell time-sharing?

Gould: Yes, it did, to people interested in doing statistics.

2.4   The arrival of Bill Rogers on the scene

Newton: Who was the next person who had an impact on Stata?

Gould: That was Bill Rogers, and there was certainly overlap between Bill and Sean. Bill was a statistician at Rand Corporation at the time. Bill is a really fine programmer. He had written a little editor. It was a very nice editor, a program called ESP. Had he marketed it right, he probably would be very rich right now, owning some giant software company that makes editors. But you know how these things go; he never dedicated his days and his nights to this editor. He came on board to help me with Stata because I wanted somebody at this point. Finis had convinced me early on of that. I could not be more pleased that he did. Now after Sean was suddenly not going to be around, I was in a panic. Finis suggested that I talk to Bill Rogers because I knew him. And I did know him; he is very approachable, very nice guy. So I did talk to him, and he was interested.

He had known about Stata early on. He'd even had a little bit of input during the very earliest developments. So he agreed, and there were some things we wanted to add to Stata. He is the person who was responsible for Stata's matrix programming language, for the first version of `ml`, factor analysis, and some other features of Stata. He knew statistics, unlike me at the time, and that worked out really well. He added the Huber stuff, robust stuff, survey stuff, whatever you want to call it, which became very important later. At that point, it was just called the Huber stuff, and nobody seemed to be very interested that we had it, or cared. There really was no market for it. He explained to me that there is a difference; that all weights are not the same. Now I understood that, but at the time, it was "Gosh, that is interesting"!

Newton: Were weights something that were involved right from the beginning?

Gould: Weights were involved right from the beginning because some economists would look at state-weighted data. That is, they would look at 50 observations, one for each of the states. Remember the first version of Stata; you really thought that you were not going to be dealing with people processing census datasets. You thought you were going to be dealing with people analyzing summary datasets. And quite often those datasets would have average income in the state, average education in the state, etc., and you'd weight by population.

Newton: So it was very natural. One of the strengths of Stata is its weighting system and the things you can do because you have those capabilities. I grew up as a SAS user, and I don't remember doing a lot of things with weights.

Gould: SAS didn't really have those extensions in the those days; nobody did. They had weights, but what they had in mind is that you had cell average data.

Newton: So now you have made several versions, and Bill Rogers has made a big contribution. Do you have a memory at what point when Stata reached a place ...

Gould: I sort of remember it in a negative way. There came a point where the financial position of CRC was not looking real good. If you sat down and started looking at the numbers, what you discovered was that Stata was carrying the mainframe, not the mainframe was carrying Stata.

Newton: Is that right? Do you remember what year that was?

Gould: That is just before we moved to West LA. We were in Santa Monica, and then we moved to West LA, and that is where the mainframe wasn't. Stata was making money, not a lot, but it was covering its cost. As long as you were used to living on a low income, and I was.

Newton: Is it fair to ask how many users there were?

Gould: Back then? I am not even sure I know. There were probably 500–1,000 users, something like that.

Newton: I promise I won't ask what it is now.

Gould: Enough to support all of this. [Gould raises his arms to indicate the building the interview was taking place in, and the second building Stata is constructing on their campus.]

Newton: At some point, you are going to move from LA to College Station, but let's talk about the language first.

Gould: Let me help you with all the moves because it gets a little confusing. Stata development began in Santa Monica. We were located on Wilshire Boulevard. We then moved to West LA, and we were there for three or four years. Then, we moved back to Santa Monica, to a 6th Street address. I guess we were there for three or four years, and then we moved to Texas.

2.5 The birth of ado-files

Newton: Let's talk about the language for a minute, and then we will talk about the move to Texas. How did ado-files come about?

Gould: By accident, that's not the official story. At the time, we talked about how brilliant we were. There was a grain of truth to the party line, but like all good things there is intent and there is accident—that come together fortuitously. That is exactly what happened here. From the beginning we knew this was important, even before Stata 1 hit the market. Stata was going to be a package that was much more open to statistical users to be developers then any other package was at that time. The way the market had developed was that the earliest packages were written by people who analyzed data; they really weren't very good programmers. But, they were awfully good at data analysis, and they wrote exactly what they needed, and they were sort of crude. And, you have BMD back then and other packages. Those packages became popular, and they spun off other companies, and then other people started working on them and making them better, and then you ended up with BMDP and with SPSS based on the BMD code base. You ended up with SAS, but that was not developed off another code base. At this point, you have really good programmers, as well as data analysts, working on these packages, but the bottom line was that if you were interested in adding a new feature to any of those packages, you had one choice: You needed to fill out an employment application to work at one of those companies, and that was it. If you wanted to see those new features, you had to beg and plead with those guys to see it, or it wasn't going to happen. In Stata, there was a `program define` statement long before we actually announced it. We were playing with it—we knew we wanted to have programmability.

Newton: Extensibility?

Gould: Extensibility—that's exactly right. Once we were satisfied with it, we went ahead and announced it. At that point, we used do-files, so you would just put together a whole kit of programs, and that is what we called them. You would have to load them, so we had a data-management system and a graph kit, and a stat kit, and you had to load one of these kits, and Stata would think awhile and then get it loaded and then you could use any of those commands that were in there. If you wanted to do data management, unfortunately, you would have to load the data-management kit because it wasn't possible to have all these things loaded at one time. We were coming up on a release, and Bill Rogers was running late, and he needed a few weeks to finish off factor analysis. I was done with everything I was supposed to have done, so I was looking for something to do. I went back to my office and thought awhile and said I could do this thing we could call automatically loaded do-files.

Newton: What platform were you doing this on?

Gould: Probably DOS, no certainly under UNIX. The first versions of Stata were developed under DOS, but once Interactive's 386/ix became available for 80286s and Sun-3s became available, I did all my development in UNIX.

Newton: But you tried to maintain the portability between these two platforms?

Gould: Absolutely right. We maintained portability. We were constantly compiling it and making sure it worked.

Newton: PC side, UNIX side, but not Apple, there was no way to do it at the time?

Gould: Bill Rogers did the first Apple port but at a later time. Once we had UNIX, we did all of our development under UNIX, and that is mostly because UNIX is a better multiuser. Computers were slow. The Sun 3 was an awfully slow computer, but I could get a lot more work done on the Sun 3 in a day than I could on a 386.

Newton: Plus it had all the tools?

Gould: It had all the tools. What did it matter that I typed `compile` and the thing had to sit there 15 minutes, when I could put an ampersand at the end of it and go on working on the next piece. Everything was just sort of easy.

Newton: Right. In this period of time ...

Gould: Oh yeah—ado-files—, so Bill says, yeah, sounds like a good idea. And neither of us had any appreciation for what a brilliant idea it would turn out to be. We just had the time, and I was looking for something to do. It sort of fit in with the direction we wanted to go. Of course, once we had it, we discovered—my goodness—all the things we could do with it.

Newton: Was it born fully formed? Did it have `syntax`?

Gould: Not `syntax`, there was an earlier form called `parse`. It had actually existed before ado-files. You have to understand, the whole programming language existed before there were ado-files because we knew we wanted that component. We just hadn't figured out that one program, one file: look on the disc, load it when you need it, and throw it away when you don't need it anymore. It was the only thing we missed, so we sort of got it backwards. If you'd thought about it right, we would have done the ado-file first and then added more programming. We had programming and then ado-files.

Newton: We are getting far afield here, but you could parse in an interactive mode in a do-file?

Gould: No, you've got to understand. Stata had `program define`, so you could add a new command to Stata. Now when is Stata going to know you have that command? Well, you put that file in a do-file. You load that file, and then Stata would have that command. When you were tired of using that command, you would erase that command because maybe you needed the memory back. All memory management was on you. All loading of the file was on you, but you could do this. So you had extensibility before there were ado-files. ado-files just made that extensibility easier to use, so we shifted a lot of our own development in that direction.

Newton: So originally almost all of Stata was written in C, and then it started moving more and more into ado-files. So what fraction now is C, and what fraction ... ?

Gould: There are two answers to that question. If you just say I am going to draw features randomly, I would guess that 85–90% of Stata is ado-files, and 10–15% is built-in. If you say, let's draw using the probability that a randomly drawn user uses the commands, you will probably find that 50–60% of it is built-in and 40–50% is ado-files.

Newton: Because most people are doing similar things most of the time and they are highly numerically intensive.

Gould: Yes.

Newton: I've enjoyed our conversations in the past about speed. Is speed as important as it used to be?

Gould: No, less important. I don't want to say it's unimportant. The faster you can do things, the better; people are doing a lot more bootstrapping and that kind of stuff. You have to look at how much faster CPUs have gotten, and you have to realize how slow the old 8088 was and the 80286. Speed was everything in those days.

Newton: What about size?

Gould: And size.

Newton: The thing had to fit on a disk.

Gould: It had to fit on a disk, it had to fit in the memory partition allocated to it. That was very difficult to make happen.

Newton: Right. So, we've got ado-files, and you are gradually adding more and more features every version. At this point, ...

Gould: Speed really stayed up, even though I said speed wasn't important, the speed kept up. Most software has developed a lot of new features that run slower, and you need a bigger CPU. Stata, every version, has a lot of new features, and it runs faster, even on the old CPU.

Newton: That was an unusual phenomenon, particularly on PCs. That was why you had to buy a PC every six months because the software got so much more extensive.

Gould: That's right. Well, we only recently, in the last two or three development cycles where we have thought, oh we can just go out and grab 16K–64K of memory. I have to remember because I still think in an old fashioned way, "Whoa, that's 64K!"

Newton: I think the same way I am afraid.

Gould: Not a bad way to think. That is what keeps Stata fast because resources aren't free. They cost something, they have just gotten hugely cheap. If you start acting like they are free, then you start generating software that runs slow.

2.6   The move to Texas

Newton: Let's now talk a little about moving to Texas.

Gould: Okay, the move to Texas was actually caused by the success of Stata. By the point we wanted to move to Texas, we knew we had finally made it. That meant that I was not going to have to change careers and go find a real job. I could plan on doing this until retirement. We were growing. We had more sales than ever. Things were actually starting to look really good. Money was coming in, but what we couldn't find were people to work for us. You could hire anyone in the LA area, but you couldn't get anyone to move to LA. No way someone was going to move from the Midwest to Los Angeles and trade in their home for a condo. It was impossible, and we were suffering from it. We had Bill, and he was wonderful, but we needed more, and we couldn't find any of these people. Technical support was even harder. You have to remember in those days I was technical support, and Bill Rogers was technical support. Whoever knew Stata or knew Statistics was technical support. We really wanted to move to a college town where we could get graduate students to be on technical support. Finis Welch (who I have previously mentioned) had by that point moved to Texas A&M University Economics Department. He of course wanted us to move to Texas. I knew that we did not want to move to Texas. I was a typical Californian, and I had this view of Texas. It was sagebrush, and flat, and sandy, and full of rednecks. That was Texas. But I agreed that we should at least look. So my wife and I came here, and we went to Houston where Finis had an office for his consulting business. We came up and looked at College Station, and I think I even met you at that time. Texas was completely different from anything that I had expected. I liked it, and the people were friendly. It was also very scary moving Stata from Los Angeles. You have to understand that LA is a huge city, and all cities look small after you've lived in LA or Manhattan. Houston may be the fourth largest city in the US, but it is small compared to LA where you can drive for an hour and a half without leaving the city. It was very scary to think about moving, but one of the things College Station had going for it is that we had a lot of users in Houston at the Medical Center. If College Station did not work out, we could shift to Houston without actually tearing the guts out of the company with another big move. That made College Station very reassuring. It seemed very small compared to Los Angeles, but it seemed safe. Finis was here, and that counted. A&M seemed friendly, so here we were. We showed up and never wanted to leave. There has never been an instant when we ever even thought about it.

Newton: So you've actually become sort of a Texan?

Gould: I guess I am a Texan. If you've lived here for more than 10 years, does it count?

Newton: I've lived here 26 years, so I guess I think so.

Gould: I have a cowboy hat.

Newton: Yeah, I guess we're Texans now.

Gould: I guess I am a Texan once I think it is okay to wear it to work.

2.7   How new Stata features happen

Newton: I remember fondly when Stata did move to College Station. It's been great for A&M, too. Let's talk about a few of the basic capabilities that I see a lot, like `ml`, bootstrapping, etc. How do such things develop?

Gould: It is not that I sit in my office and I have some plan, I see a vision out there and I say "yeah that's where we are going". At any one time, I think I know where we are going, but I can tell you I have never been right yet. What you try to do is find a direction to go. It's like a hill climb—it's just like a maximizer: "This looks like a good way to go, but don't go too long before you evaluate the derivatives again." One of the things that has really worked well for Stata is things like Statalist, the *Stata Journal*, the Users meetings that were started in London, that are now going on in Boston and Berlin, and other places. We get a lot of user feedback, but talk is cheap. Someone calls you up and says, "I want to see this." If we implemented every time someone calls us up with "I want to see this", it would be ridiculous, and you couldn't do it. But we actually get some real indication because Stata is programmable; we are able to go to the Users meetings; we can see that people are writing programs to do such and such. That is what we pay attention to. We say look where they are going, and then we follow. We see where the users who are active are going, and then we lead from that direction and say, "Follow us!"

Newton: That is one of the great things about extensibility.

2.8   The Stata Journal

Newton: Once the Internet sort of made the *Stata Technical Bulletin* not as badly needed, a group of us got together in Boston and said, "Let's create a journal."

Gould: I think that was really spawned a lot by Nick Cox. In part, StataCorp's concern was what was going on with the *Stata Technical Bulletin* because we knew it wasn't working as well as it used to work. Remember the *Technical Bulletin* was primarily a means of distributing user-written commands.

Newton: Well, it wasn't as needed?

Gould: Right.

Newton: Many small articles describing new commands weren't as needed because you could do that on the Internet.

Gould: Or in the Boston archive.

Newton: So several people got together in Boston and said, "Wouldn't it be great to have a journal?" And you agreed to do that.

Gould: Again this is once more—you know how some restaurants are run where they have a window onto the kitchen? They do that for a couple of reasons. One is that it is sort of entertaining for the diners to be able to look into the kitchen, and the other reason is that it keeps the kitchen clean because those people know they are being looked at and it sort of monitors itself. That is exactly what we have tried to do with Stata in terms of the *Stata Technical Bulletin*, Statalist, and now the *Stata Journal*. You notice we have always put them under independent management. We don't decide, we put up with an independent group deciding. We observe what they do, and sometimes they surprise us and go in directions we did not anticipate them going. Had we been in control, we would never have let it happen because we would have stopped it early on and said that is the wrong direction. Look what they are doing! Like the Boston Archive. The `net` command worked out wonderfully but didn't work out like I expected. I thought there would be no big single place, that it would be scattered all over the place and designed the command to work in that environment. Well, it turns out that up pops the Boston Archive, and most things go there. Fine! I didn't anticipate it. I was wrong not to anticipate it, but I wasn't in control of Statalist where I could say, "No, no, no, Kit [Baum], that's not the direction we are going." Kit was absolutely free to get out there and say, "I have built this thing, and here it is." I didn't think it was the direction that ultimately we would end up going, but I was wrong. This is not me being good; it is not a moral position, it is actually in our interest to let the users go. Again, leading from behind and sitting back and watching them. See what they want to do, and then we try to catch up.

2.9 Modern StataCorp

Newton: One thing I have always noticed about Stata is that you don't have hordes of programmers here. Is that a fair statement?

Gould: You make us sound small!

Newton: Of course not! You are not small, but you personally write some of the code.

Gould: I am a big believer in using very competent and senior level developers and giving them a lot of leeway, and expecting a lot of them, rather than a bunch of coders executing instructions from on high.

There was a time when I was Stata, but that time has come and gone. I get interested in a project, and I work on it, and I have great fun doing it, but now there are a lot of people working on Stata.

Newton: And you spend the rest of time on business?

Gould: About 50 percent. Finis Welch has been the invisible partner in this from the beginning, and this is the person I talk to a lot. Stata would not be here but for Finis Welch. I could have put this company out of business 15 times between start up and where we are now if somebody had just given me the money with which to do it. There is a lot of learning by doing. Finis has been very good all along in letting me make my own mistakes but keeping me from making mistakes that are too big.

Newton: Finally, what can you tell us about the future of Stata?

Gould: Like I said, we do not set long-term goals, at least with the expectation of keeping them. We say where it is that we think we want to go, and then we work out what we ought to be doing for the next one or two years. By that time, the long-term goal will have shifted, and the process repeats.

Nevertheless, we are committed to statistics. This is what we are. This is what we do. We are committed to the professional research market. We are not going to make a stab at telling the academics and researchers to go away so that we can focus on the Fortune 500 and turn Stata into a system for use by people who do not know statistics. We assume that the person behind the keyboard knows more than we do, not less, and that is the market we are going to continue to serve, without question.

Newton: Thank you Bill. I've certainly enjoyed this conversation, and I look forward to seeing all the exciting things in Stata's future.

3 In at the creation

Sean Becketti

I got sucked into helping launch Stata because my TV died.

I was an assistant professor at UCLA, trying to raise two small children during a governor-mandated salary freeze at the University of California. To make ends meet, I did some consulting, but the consulting was cutting into my research time and hurting my chances for tenure. I finally vowed to stop consulting completely, even if it meant economizing at home. Then my TV died.

It was a hand-me-down, a dinosaur, a massive (for the time) 25-inch console color TV, almost impossible to move. It had been ill for a while—most notably losing all video during a touchdown run in one of the Super Bowls—but I nursed it along as best as I could until it finally failed completely. An infant and a toddler at home, no money for babysitters and going out, and no TV: not a recipe for a happy home.

At that moment, one of my colleagues at UCLA asked if I would work with Bill Gould on a scheme he had for a statistics package that would run on the IBM PC. I met Bill when, as a graduate student, I was briefly assigned to him as a research assistant. Later I worked with him on a consulting contract. At the time, I do not think either of us was particularly impressed with the other. I believe Bill thought I was a bit of a slacker; hiding from him when I was his research assistant did not make a good impression. In return, I thought Bill was a bit eccentric. I thought that the idea of writing a serious statistics package for the IBM PC of 1984 seemed daft. The CPUs were painfully slow, there was not much memory, and hard drives had not been introduced yet.

The idea that I should help with this project was not entirely random. I had paid part of my way through graduate school by working on a successful mainframe statistics package, and I had contributed a few nice features to it. The notion was that I would look over Bill's shoulder, test the early versions of the package, and offer some advice on one or two econometric techniques that we thought were important for marketability. It seemed like a waste of my time, but it also seemed like a quick and easy way to earn enough money for a new TV.

I was pretty cocky at the start. After all, I had worked on a "serious" commercial mainframe package. And many of Bill's ideas mystified me. I let him babble on about preparsing and strict syntax and importing cross-product matrices. I did not see the point of most of it, but Bill seemed very enthusiastic about the program.

Anyway, I had fun testing (that is, breaking) the software. At first, I was able to quickly generate crashes or serious errors in each new version. As time went on, though, the program became more and more reliable, and I had to exercise considerable ingenuity to unearth a significant problem. Moreover, the longer I worked with Stata, the more I began to appreciate Bill's design. I started to see that my notions of interface design, learned in a card-reader-oriented mainframe environment, were clumsy compared with the interactive interface Bill envisioned. I began to appreciate the modeless nature of Stata. While Stata's initial list of features was puny compared with established packages like SAS and SPSS, Stata was a lot more fun to drive, with no ping-ponging between DATA steps and PROCs. Stata seemed to facilitate a sort of dialogue with the data. Each command was a question asked of the data. Each result suggested further questions, and Stata made it easy to pursue these questions. I did not have to modify a program, resubmit it, and wait for the output. I just kept asking more questions until I had the answers I sought.

This interaction was a revelation, and it changed my approach to data analysis. (Those of you old enough to have made the transition from mainframes to the early PCs may remember a similar reaction the first time someone showed you a simple spreadsheet in action. The immediacy was breathtaking.) Previously, I had dismissed exploratory data analysis, believing it was just a touchy-feely distraction from more definitive confirmatory techniques. Besides, the increasing availability of high-powered computers and professional statistical software (at least in the university) seemed to make Tukey's paper-and-pencil approach superfluous. Ironically, the ease of using Stata for exploratory data analysis opened my eyes to the importance of these techniques and to the importance of robust and resistant estimators in confirmatory analysis.

Needless to say, I kept working with Bill long after I bought the TV. Bill and I wrote the first Stata manual—formatted on the mainframe in the wee hours of the morning— and released the program in December 1984 at a large meeting of economists in Dallas. We returned somewhat chastened. Several other bright people had the same idea of porting professional statistics to the PC, and we realized that we had some catching up to do. (Bill received a lot of well-meaning advice to throw in the towel before he lost too much money. The consensus was that the market was already too crowded and that the inevitable entry of packages like SAS and SPSS would crush smaller competitors. Fortunately, Bill is very stubborn.)

The next couple of years were among the most stimulating of my professional life. Stata grew quickly in power and features (although sales were slow at first), and each addition to the program posed tough puzzles for us to solve. But beyond the intellectual challenge, there was the fun of working with Bill. Perhaps the best way to convey the experience is to describe a typical day.

At some point, I left UCLA and moved to the Midwest, but I flew to Los Angeles regularly to put in week-long development pushes, and when I did, I was a guest at Bill's house. We rose early and opened the office. Not all the work was software design. The company was small, so sometimes Bill and I started the day by setting up PCs or assembling tables for them to sit on. Eventually, though, we would address the current development challenges. Then the yelling would begin.

Bill and I disagreed about almost everything, and we expressed ourselves colorfully and at top volume. (I remember, in particular, some arguments about hypothesis testing in the linear model.) I would suggest a possible approach. Bill would reply that the approach was obviously impossible and that only a seriously impaired intellect would suggest it. I would reply in the same vein. We were so loud and so heated that I think we scared the rest of the staff. At least, they did not interrupt us very often.

There was never anything personal in our arguments. Bill and I were simply passionate about getting the best possible solutions into Stata. If I could prove my point (something that did not happen often enough for my ego), Bill would stop dead and say, "Well, I guess I'm wrong", and things would go on smoothly until the next disagreement. More frequently, I would see the flaw in my approach and come around to Bill's point of view. At the end of the day, the arguments would stop and we would drive companionably to Bill's house, where he would prepare a gourmet dinner accompanied by some truly fine wines. (My recommendation: if Bill ever asks you to dinner, accept. He is an excellent cook.)

My professional association with Stata lasted about ten years, but my involvement as an end user has never slackened. The meandering path of my career has given me the opportunity to build research teams in several large financial services firms, and each of those efforts has required a substantial amount of proprietary model and software development. Stata has played a key role in model estimation, testing, and error tracking. I have also used Stata heavily as a scripting language, and I have built several automated reporting packages with Stata as the lynchpin. Whenever I start a new research team, I order Stata for the entire staff, but I do not require them to use it. Instead, when they bring a research problem to me, I load their data into Stata and explore the issue interactively with them. Usually only one or two exposures are required before my staff are clamoring for Stata.

I like to think that I contributed one or two decent ideas to Stata. Typically, the best ideas were not solo creations, but rather the result of discussions with Bill. The idea for the first **parse** command arose when Bill and I were struggling to figure out ways to overcome some of the limitations of do-files. I think the **pause** command resulted when I was stymied when trying to develop a more primitive version of the concept and I asked Bill for help. He realized that it was a tougher problem than it first appeared, so he then developed the more elegant and useful **pause** command. Actually, many Stata users around the world have contributed to the evolution of Stata. I had the privilege of meeting (at least electronically) some of these contributors and previewing their enhancements when I was editor of the *Stata Technical Bulletin*.

Being present at the birth of Stata was a stroke of good luck for me. The creative challenge was intensely satisfying intellectually, and working with and getting to know Bill was a high point personally. And I did get a 19-inch RCA color TV.

Speaking of TVs, my wife has been lobbying for a high-end flat panel HDTV with a state-of-the-art surround sound system. Maybe I will give Bill a call.

Part II

A vision from outside

4 Then and now

Sean Becketti

4.1 Introduction

I have used Stata for a long time. A very long time. Since before it was released. Since before it was called Stata.

As I described in an earlier article (Becketti 2015), I had the good fortune to work with Bill Gould as he was developing what would become Stata. My initial job was to test—that is, to break—early versions of the code. As the program matured and breaking it became more difficult, I spent more time talking with Bill about what Stata should do and how it should do it. I have designed a few pieces of my own software over the years, but none of that was as much fun as the discussions Bill and I had in those early days about Stata.

In the first couple of years of Stata's life, I remained involved in the discussions about its evolution. Over time, though, my professional responsibilities led me in different directions. I do not remember exactly when the balance tipped. I remember working with Bill on the first version of the Stata `graph` command (which I still love) and on the `anova` command. I also recall assisting with improving the matrix inversion algorithm that underlies the `regress` command. That project led to Bill's first prototype of a matrix language for Stata. Anyway, at some point around then, I stopped working on Stata. A bit later, Bill asked me to edit the *Stata Technical Bulletin* for a couple of years. However, my professional obligations eventually took over all the time I had, and I cut the cord. I did return to the fold temporarily. Thanks to the global financial crisis, I had some free time in 2009, and Bill persuaded me to write *Introduction to Time Series Using Stata* (Becketti 2013) for the Stata Press. As it happened, I landed a new job fairly quickly, and I ended up trying Bill's patience by taking an unconscionably long time to finish what is a fairly straightforward book.

All of this is to establish that I have been involved with Stata for its entire life. Stata has changed quite a bit from the pre-Stata I first tested in 1984 to the software system it is today. The ways I use Stata have changed as well.[1] I will spend the next few pages highlighting some of the significant milestones along the way—for Stata, for time series, and for modeling in financial services (my field of endeavor).

---

1. Surprisingly, I have remained exactly the same as I was in 1984.

25

4.2   Stata, then and now

Stata launched just about the time everyone else in the world decided to introduce a statistics program for personal computers. PCs were weak; Bill started developing Stata before hard disks were offered on PCs. They had just become powerful enough to consider moving some real work from the mainframe (which was expensive to use) to the PC (which was expensive to buy, but free to use thereafter).

Many colleagues and competitors tried to convince Bill that he was introducing his "nice little regression program" (their description) too late—the market was already crowded with programs that had many more features than Stata 1.0. In those days, potential customers were fixated on comparative lists of available features (even when they had no plans to use many of these features). Did Stata include analysis of variance? Not yet. Did it include time-series analysis? Not yet. Did it include graphics? Too few to tout. To make matters worse, many customers were waiting for the two behemoths of statistical software—SAS and SPSS—to offer PC versions and crush all the upstarts. What these observers missed were Stata's considerable advantages.

Three things set Stata 1.0 ahead of the competition, more than making up for Stata's initial paucity of statistical features.

Stata's first advantage is its "modeless" approach. In many competing programs, the user has to navigate from module to module to get anything done. The user has to run a data-preparation step, then run another module to estimate a model, then save the estimation results, and then, finally, generate predictions and diagnostics. Each step requires exiting one module and starting another, and results that suggest alternative models require the user to start the process over from the beginning.

In contrast, all functions of Stata are available at all times. To do something in Stata, you—as the Nike slogan goes—Just Do It. As a result, Stata users typically do a better, more thorough job of exploring all aspects of their data than they would if they were using equally powerful, but more cumbersome, tools.

The second advantage, closely related to the "modeless" approach, is the data rectangle (my term, not Stata's). Other programs operate on files stored on disks. In data-preparation steps, other programs retrieve one observation at a time from an input file, manipulate it, and then store the results in an output file—again one observation at a time. Then the output file is passed to one of the statistical modules. This file-based approach dictates the module-to-module processing described above.

As you know, Stata stores its data in memory in a rectangle.[2] Each column contains all the observations on one variable. Each row contains one observation of every variable. For spreadsheet users, this is a natural, intuitive structure. In addition, it is easy to use *varlists* and the `if` and `in` clauses in Stata to restrict the operation of a command to any desired subrectangle without having to actually subset the data. It is difficult to overstate the usefulness of this approach.

---

2. At least, you can visualize it as a rectangle.

Initially, the data rectangle was regarded as a disadvantage of Stata relative to its competitors. After all, only a limited amount of data can fit in the computer's memory, and in the early days, that amount was substantially less than it is today. With the expansion of computer capacity and simultaneous reduction in cost, there are relatively few problems that do not fit in memory today. The data rectangle is now a clear advantage, and file-based approaches remain clumsy in comparison.

The third advantage is Stata's rigid syntax. Things have loosened up slightly over time, but initially there was only one way to say anything in Stata,[3]

> by *varlist*: *command varlist* = *exp* if *exp* in *range* [*weight*], *options*

where everything except the command name is potentially optional. This may not seem like an advantage or a disadvantage, or even anything worth mentioning. But, while this syntax might seem confining at first, it actually simplifies learning Stata. If there is only one way of saying things, then you have to learn only one way of saying things. As a result, it is usually easy for users to guess how to use new commands. Even guessing what a Stata command might be called (is there a common English verb that describes this action? `regress`? `summarize`? `describe`?) works surprisingly often. This simplicity drastically shortens the learning curve for Stata.[4]

As I mentioned before, early users were "feature freaks". The important things—modeless operations, data rectangle, consistent syntax—are not immediately appealing, and their benefits are obvious only after one uses Stata for a while. As a result, sales grew slowly at first. Eventually, though, loyal users started spreading the word. Moreover, Stata users started inventing Stata commands, expanding the range of things Stata could do conveniently.

This brings me to the next—and, to my mind, most important—milestone in Stata's progress: the ability to program Stata in Stata. It was always possible to create a Stata script to package a long or complex or just plain tedious sequence of operations.[5] A Stata program—a script that appeared to the user to be just another Stata command—took longer to appear. But once it arrived, the focus of Stata development changed.

Being a lazy sort, I frequently would ask Bill to add a feature or command to Stata that would make my life easier. Bill would counter that a resourceful person (clearly not me) could produce the requested result with a user-written Stata program, obviating the need to clutter Stata with a new feature. Then things got competitive. I would try to prove that it was impossible to fulfill my request with existing Stata features.

---

3. I am fudging the facts ever-so-slightly. For example, the `set` command.

4. This syntax did not always make things easy for us when Stata was being invented. We struggled first to pick straightforward, descriptive names for the commands (as you can tell from the current manuals, this became harder over time), then we struggled to figure out how to translate a natural English command (perform this type of estimation in the following way with certain adjustments that are very important to us) into the Stata syntax. More than once, we stared at the keyboard together for 10 or 15 minutes to see if there was an unused special character that might simplify things.

5. I do not remember exactly when the `do` command was introduced, but I do not believe it was available in Stata 1.0. Nonetheless, there were work-arounds that offered similar functionality. Anyway, the `do` command was introduced very early.

Bill would try to show how, in fact, my task could be done with Stata. On those rare occasions where it turned out I was correct, Bill would add another programming construct (looping, macros, etc.) to the Stata program toolkit, and that would solve the problem.

As we tackled more and more complex tasks, the Stata programs became more complex, and a serious problem arose. Stata programs were often difficult to write—not because the calculations were complicated, but because it was tedious to interpret the user's command line. And, more troubling still, Stata programs violated the syntax introduced in Stata 1.0. Abbreviation of variable and option names typically could not be supported. Indeed, options usually could not be supported, at least in the form used by built-in official Stata commands. Stata programs were vulnerable to simple user typos in ways that built-in commands were not.

And then a miracle arrived—the `syntax` command. Now, with almost no effort on the part of the programmer, a Stata program could be virtually indistinguishable from a built-in command. Typos triggered exactly the same error handling and error messages produced by a built-in command. Stata programs became much shorter and easier to read. I believe the `syntax` command was the catalyst for the explosion of Stata programs—and, hence, the explosion of Stata features—that followed. Other commands (`set trace on/off`, `pause`,[6] extended macro commands and the like) have helped, but `syntax` is the secret ingredient. The Stata programming features available today make Stata one of the most efficient platforms for complex statistical program development.

At this point, you are probably thinking, "What is wrong with this guy? The `syntax` command is so important? Come on. What about Stata's incredible range of statistical capabilities? What about Mata? What about Stata graphics and the Graph Editor? Are not those the signature features of Stata?" Of course all of those things and more that I did not mention are important, and they are typically the reason many users start using Stata. Many other programs offer a broad range of statistical features and graphics and even matrix languages, but they are not Stata. I have used many other programs, and many of them are very good, but none of them quite capture Stata's balance between already-built-in features and the convenience of extending and reusing these built-in tools in new and creative ways.

I would be remiss, however, if I did not point out the Stata team's absolute commitment to statistical integrity and accuracy. Those of us who survived the early days of statistical software remember some of the serious shortcomings of many of the early commercial programs. Working statisticians developed Stata. They insisted on a high-quality, no-compromises program to use in their own research, and they have firm opinions about the right way to do statistics. Of course, at times, the market demanded that Stata include a statistical technique the Stata team would prefer on professional

---

6. I believe my constant complaining played a small part in Bill's invention of the `pause` command. One could argue that my primary contribution to Stata has been a steady litany of (constructive) complaints. If StataCorp ever creates a position for a "complainer-in-chief", I think I have got a good chance of getting the job.

grounds to omit.[7] Even in those instances, the developers insisted on implementing the highest-possible-quality version of the technique.

Before I turn to the evolution of time-series analysis over the last 30 years, I want to say a few words about the implementation of time series in Stata. It took a long time to get built-in time-series capabilities in Stata. For many years, I relied on my own home-brew Stata programs to handle my time-series analysis needs. When Stata finally introduced its time-series suite, I was reticent to leave my familiar ado-files behind. My programs retained the modeless approach of Stata. In contrast, the Stata time-series tools required me first to define the frequency of the data and the variable that contained the date and time information. Many of the time-series commands used the "two-part" syntax that Stata has adopted for features that are too complicated to squeeze into the Stata 1.0 syntax. For example, to estimate and apply an exponentially-weighted moving-average smoother—a bread-and-butter operation in the world of time series—you cannot just type, say, `ewma`. You have to type `tssmooth exponential`, and the `tssmooth` command includes a family of related smoothers that are specified by the second word of the command. This syntax works, but it lacks the simplicity and elegance of equally sophisticated cross-section commands (for example, `regress`).

I held out for quite a while, struggling along with my more-limited ado-files. Over time, though, I found that my staff readily adopted the Stata time-series suite, and soon they were able to do easily what took me a lot of work. I started using one or two of the time-series commands that I just could not live without, but I did not transition completely until I wrote *Introduction to Time Series Using Stata* (Becketti 2013). That experience forced me to catch up with the rest of Stata users, and I am glad I made the change. I believe the Stata design makes time-series analysis about as easy as it can be. The complexity of the time-series commands reflects the complexity of the statistical techniques.

4.3 Time-series analysis, then and now

There is an element of irony in talking about changes over time in time-series analysis.

In the last 30 years, there have been a remarkable number of major breakthroughs in time-series analysis.[8] When I began my professional career, there was a handful of techniques for pure time-series forecasting; a fair amount of attention to seasonal adjustment; and some careful thought, but few tools, devoted to analyzing business cycles. Autocorrelation in the error term was regarded as a particularly annoying variety of heteroskedasticity that required "correction".

Box and Jenkins reinvigorated research in time-series analysis by providing a unified, tractable, and intellectually satisfying perspective on dynamic analysis. Just as

---

7. To avoid giving offense, I will not name any examples.

8. Warning: In this section, I am going to ignore a lot of interesting aspects of time-series analysis. I come to the field as an applied economist with a strong preference for time-domain models. Extensive work has taken place on frequency-domain models and in related fields that use time-series techniques. I am limited by my range of experience.

important, in their 1970 book, Box and Jenkins published usable algorithms—recipes for time-series software that put these techniques within reach of most statisticians.[9]

Then the pace of progress accelerated. Hendry and Mizon (1978, 1980) inverted the view of autocorrelation as a problem to be corrected and focused attention on the substantive information in a dynamic specification. Engle's (1982, 1987) introduction of autoregressive conditional heteroskedasticity gave researchers a way to model the impact of transient disruptions to dynamic processes. Granger's (1969, 1987) concept of cointegration provided a better way of thinking about long-run relationships. Sims's (1980) analysis of vector autoregressions challenged traditional thinking about structural econometric models. At least three Nobel prizes recognize the importance of this work and several more reward work that draws upon these contributions.

In some ways, though, the central topics in time-series analysis have hardly changed.

- Dynamic models of stationary variables provide a useful but dim flashlight into a dark future. They improve our ability to gauge the impact of transitory disturbances but, by construction, their predictive power decays rapidly.

- Small differences in a trend make a big difference in long-run outcomes. For instance, a difference of half a percentage point in the annual rate of gross domestic product growth generates a gap of ten percentage points in the cumulative growth over 20 years. Unfortunately, it is difficult to identify trends precisely or to distinguish them from unit roots.

- An irreducible tension remains in time-series analysis between pure forecasting models such as vector autoregressions and structural models. Sims offered vector autoregressions as a way to break free of the often-implausible simplifying assumptions in structural econometric models. But, without some structural assumptions, policy analysis—the ability to predict the outcome of counterfactual interventions—is impossible.

The statistical advances of the last 30 years have improved our understanding of these challenges, but the challenges remain.

4.4   Rocket science, then and now

For the last couple of decades, I have worked in and led modeling and analytics teams in financial services firms. I was lured into this line of work by newspaper and magazine stories about the glamorous lives of the Wall Street "rocket scientists"—mathematicians and statisticians who unlocked the secrets of unlimited wealth.[10] While I am not averse to unlimited wealth (I am still looking for it), I was also attracted to the opportunity to use my econometrics skills to do more than study the real world. I wanted to change the world.

---

9. In those days, most researchers had to write their own programs to use new techniques.
10. There may have been a touch of hyperbole in these accounts.

The reality I encountered as a modeler in industry was quite different from the impression I had from media accounts. It would take more space than I have here to describe the challenges of offering technical expertise to an organization full of "practical" executives. While there are many brilliant people in financial services, it is difficult to do justice to the depth of misunderstanding of statistics and models in industry—some of it willful, some of it honest ignorance, and some of it due to a poor ability on the part of the "rocket scientists" to explain what they are doing. This lack of understanding played a significant role in the global financial volcano that erupted in late 2008. As important and interesting as these issues are, I am going to set them aside and talk instead about the change in the pre- and postcrisis roles of models and modelers in financial services.

Two true stories will illustrate the precrisis environment.

My first industry job was on Wall Street. I built models that predicted how many people were expected to prepay their mortgages each month. Prepayments are difficult to predict accurately, and small improvements in accuracy have a large impact on financial outcomes. I worked on models for the mortgage-backed securities issued by Fannie Mae and Freddie Mac. Next to me sat a gentleman who worked on models for Government National Mortgage Association (GNMA) mortgage-backed securities. One Friday, he finally finished his models and, as a reward, he was immediately fired.[11] The next Monday morning, our information technology manager asked if he should put the GNMA models into production, making them immediately visible to the global trading and sales staff. Although the ex-modeler and I were friendly, he worked for a rival "clan" in our firm, so, in accordance with the Wall Street code, we never shared information about our models. I had no idea what was in his GNMA models, and I told the information technology manager so. His response: "You are the prepayment guy now. Do I turn these models on? Yes or No?"[12] That was the extent of change control and model risk management.

---

11. I made a mental note to dawdle over my Fannie Mae and Freddie Mac models.
12. I gave the correct answer.

Some years later, I accepted a job at a mega bank to build a research department and create an internally developed suite of mortgage models. However, before we developed our own models, I was asked to repair an existing model. Suffice it to say that this existing model fell far short of even the shoddiest of professional standards. The bank employed this model to value and hedge many millions of dollars in mortgage assets, and the model was broken, so I reverse-engineered it over a weekend. On Monday morning, I showed the first test results of the model to the trading floor and the head of capital markets. I cautioned them that this was the first test run of the new code. They looked at the results, looked at me, and then pointed out that the model was indicating that they needed to add a staggering amount of assets to balance their hedge appropriately. I responded that yes, that's what the initial test results indicated, but I cautioned them once again that this was a test (and I hinted that the model was not the most reliable). There was a moment of silence, then the head trader said, "Let's do it". The traders then all filed out, went to their desks, and picked up their phones.[13]

The modeler's life is very different in the postcrisis world. The rules for change control, model risk management, and model validation are many and stringent. I still lead modeling groups, but I spend the majority of my time with regulators, internal auditors, external auditors, model validators, legal and compliance staff, and the like. In addition to the extensive reviews and controls surrounding models, there are specific directives from the regulators on how to properly build models. Some of these directives are a useful corrective to the excesses of the past. However, in my opinion, some of these directives also reveal a poor understanding of statistics and a troubling faith that risk can be eliminated through tightly controlled modeling practices. Again, this is a big topic worthy of a book-length treatment,[14] but it seems clear to me that we are not yet in equilibrium.

Practices clearly were very loose in the precrisis days, but for the most part, there was a simple but effective control. If your models helped the firm make money, you kept your job and got paid. If your models led the firm down the wrong path, you would be looking for a new job. You had a strong incentive as a modeler to do the right thing. Sadly, in the global financial crisis, millions of people who have nothing to do with the financial services industry lost money, lost jobs, and lost homes. The rules, understandably, have to change.

As much as it complicates my life, I understand the need to improve controls around models in financial services. My concern is that these controls increase the cost and reduce the nimbleness of financial services without actually preventing future financial missteps. There is not sufficient space here to make this argument, so I will content myself with one observation.

In the savings and loan crisis, 1,043 out 3,234 thrifts closed between 1986 and 1995. While fraud produced some of these failures, a critical factor in the crisis was the sys-

---

13. One might conclude that the traders simply were having fun at the expense of the new guy. I lean toward this interpretation. However, there never was a review of my work and the reengineered model immediately went into production.

14. Yes, I am thinking about it. Do not get too anxious though. I am very slow.

tematic failure of most thrifts to hedge their exposure to interest rate risk.[15] In the wake of the crisis, regulators began to require regulated financial entities to implement disciplined interest-rate risk hedging. Although, as it happens, interest rate risk played virtually no role in the global financial crisis of recent years. One interpretation: regulators succeeded in protecting the system from interest rate risk. An alternative interpretation: regulators, like generals, were still fighting the last war and failed to understand the growing threat of the housing bubble or, at the very least, failed to take appropriate action.

4.5   Plus ça change . . .

Undeniably, much has changed—and much of it for the better—in the last 30 years. But the big issues are largely the same. Amazing improvements have occurred in time-series analysis, yet we struggle with the same foundational questions. In regulating the use of complex statistical models in financial services, we are still searching for ways to limit systemic risk to the global economy without crippling an essential industry in a modern economy. I suspect that 30 years from now, we will find that these issues will continue to challenge us.[16]

In the same way that the advances in time-series analysis have improved our understanding of forecasting, policy analysis, and the like, without eliminating the central intellectual challenge, Stata (and, to be fair, some of its competitors) has improved our ability to analyze, visualize, and model data. Today, we can easily do a multitude of things that were arduous or impossible 30 years ago. It is as though we have been exploring a cave, and the improvements in Stata and our statistical tools have given us a stronger flashlight. We now can see features deeper in the cavern, and the flashlight has revealed another turn in the passage that we cannot quite see around yet. Maybe in the next 30 years.

References

Becketti, S. 2013. *Introduction to Time Series Using Stata*. College Station, TX: Stata Press.

————. 2015. In at the creation. In *Thirty Years with Stata: A Retrospective*, ed. E. Pinzon, 19–22. College Station, TX: Stata Press.

Box, G. E. P., and G. M. Jenkins. 1970. *Time Series Analysis Forecasting and Control*. San Francisco: Holden-Day.

Engle, R. F. 1982. Autoregressive conditional heteroscedasticity with estimates of the variance of United Kingdom inflation. *Econometrica* 50: 987–1007.

---

15. Or it was a result of their failure to even understand that they were exposed to interest rate risk.
16. Actually, they will challenge you, not me. Thirty years from now, I am unlikely to take an active interest. (I am also sure that 30 years from now, the Cubs still will have not won the World Series.)

Engle, R. F., and C. W. J. Granger. 1987. Co-integration and error correction: Representation, estimation, and testing. *Econometrica* 55: 251–276.

Granger, C. W. J. 1969. Investigating causal relations by econometric models and cross-spectral methods. *Econometrica* 37: 424–438.

Hendry, D. F., and G. E. Mizon. 1978. Serial correlation as a convenient simplification, not a nuisance: A comment on a study of the demand for money by the Bank of England. *Economic Journal* 88: 549–563.

Mizon, G. E., and D. F. Hendry. 1980. An empirical application and Monte Carlo analysis of tests of dynamic specification. *Review of Economic Studies* 47: 21–45.

Sims, C. A. 1980. Macroeconomics and reality. *Econometrica* 48: 1–48.

4.6   About the author

Sean Becketti is a financial industry veteran with three decades of experience in academics, government, and private industry. Over the last two decades, Becketti has led proprietary research teams at several leading financial firms, in which he has been responsible for the models underlying the valuation, hedging, and relative value analysis of some of the largest fixed-income portfolios in the world.

5   25 years of Statistics with Stata

Lawrence Hamilton
Department of Sociology
University of New Hampshire
Durham, NH

5.1   Introduction

Ten years ago on Stata's 20th birthday, I told a story about writing the first out-of-house Stata book—*Statistics with Stata*, initially published in 1990b for Stata 2. My 20th-birthday story appeared in a celebratory issue of the *Stata Journal* (2005). Since then, both Stata and *Statistics with Stata* have undergone many revisions. Today, on the occasion of Stata's 30th birthday and 25 years of *Statistics with Stata*, where have we traveled? This note is a sequel.

5.2   A brief history

From my first encounter in the 1980s, at the dawn of the desktop revolution, Stata impressed me with its smooth integration of data management, regression, and graphics. These strengths suited my research and fit the graduate-level social science statistics courses I was teaching. Dissatisfied by existing textbooks and too young to know better, I decided to write my own text that would also cover Stata. This plan worked poorly because its scope proved too large. Through five years of writing, my one book fissioned into three: *Statistics with Stata* plus two full-length texts (*Modern Data Analysis* [Hamilton 1990a] and *Regression with Graphics* [Hamilton 1992]). Originally, *Statistics with Stata* was meant to be just a supplement.

My coverage dilemma soon became worse, as the famously expandable Stata program added features faster than any text could encompass. In the next edition, *Statistics with Stata (Updated for Version 3)* (Hamilton 1993), it was no longer necessary to introduce readers to Microsoft disk operating system. We were deeper now into the desktop revolution, so culture shock at the C:\> prompt had subsided. However, any space savings were more than offset by the need for new material on data management, graphics, regression diagnostics, curve fitting, robust regression, logistic regression, and principal components or factor analysis. Using smaller fonts decreased the page count; but even so, I had to omit a lot of information. I saw where this was heading; *Statistics with Stata* would need to stand alone, because there was no way that any text could keep pace.

The next edition, *Statistics with Stata (Updated for Version 5)* (Hamilton 1998), continued in this vein, with new chapters on survival analysis and programming. Likewise for *Statistics with Stata (Updated for Version 7)* (Hamilton 2002) and *Statistics with Stata (Updated for Version 8)* (Hamilton 2004), which added material on time series, matrix programming, survey data, panel data, and generalized linear modeling— bringing us to Stata's 20th birthday. On that birthday, I described Stata 8 as "the most radical upgrade in Stata's history, led by Stata's new menu system or graphical user interface, and completely redesigned graphing capabilities". Stata's innovations resulted in sweeping changes to my book as well. Graphics became the longest chapter. Stata's many reference manuals had gone virtual along with many other resources, so *Statistics with Stata* increasingly could point to those "for more information". Virtual reliance became pronounced with *Statistics with Stata (Updated for Version 10)* (Hamilton 2009), as I added chapters on survey data analysis and mixed-effects modeling, along with new sections on multivariate time-series models, Mata programming, graph editing, creative graphing, and missing values. *Statistics with Stata (Updated for Version 12)* (Hamilton 2013) reorganized much of this content while updating and adding still more methods, notably structural equation modeling. It weighed in at 473 pages.

5.3   Examples in Statistics with Stata

Through all its iterations, *Statistics with Stata* has aimed to help grad students and practicing researchers by filling the gap between Stata's vast references and some common challenges in learning and applying it to data. Filling this gap required not only highlighting certain tools and concepts, but following through with interpretation. With Stata, any fool can fit a hundred kinds of regression. But which kinds answer broad research questions? And what can one write about his or her results? Such questions recur often in conversations with students and colleagues.

To make interpretation worth the effort, I sought examples for all editions that used data with a real and consequential story to tell, such as the Challenger space shuttle disaster. Most of my choices were far less compelling, but I tried to replace examples with newer ones over time. Early editions used some water-crisis datasets, drawing on studies done in the 1980s. After finishing *Regression with Graphics* in 1992, the third book in my trilogy (and largely built around a water-crisis example), I felt burned out on writing about statistics and jumped at a chance to take these tools into the field—literally into the Arctic. My first Arctic article featured some oddball Stata graphics constructed with STAGE, the Stata Graphics Editor, a makeshift separate program that foreshadowed today's integral Graph Editor. The article also contained a photo of high school students in the remote village of Kiana, Alaska, using Stata on my laptop computer to query data from a regional survey that Carole Seyfrit and I had just conducted (Hamilton and Seyfrit 1993b). I soon included such northern examples in *Statistics with Stata*, my favorite being an illustration of multinomial logit regression that I wrote while looking out across the windswept ice of Kotzebue Sound during an Arctic winter storm (Hamilton and Seyfrit 1993a). In later *Statistics with Stata* editions, I added new surveys along with climate-change examples reflecting my contacts with physical scientists working in the Arctic.

In figure 5.1, two images from the 2013 *Statistics with Stata (Updated for Version 12)* illustrate the new climate interest and longstanding quest for examples that tell a consequential story. The four panels of figure 5.1a depict ordinary regressions (**regress**) of global surface temperature on lagged values of one, two, three, and four predictors; each predictor represents a major physical process or force known to affect global climate. Volcanic aerosols, which cool the surface by blocking sunlight, alone have short-term impacts but do poorly in predicting the overall temperature record. Volcanic aerosols plus solar irradiance performs slightly better. Although sunlight warms the Earth and undergoes 11-year cycles, the variation in irradiance is relatively small (about a tenth of one percent) and buffered by ocean heat capacity, so these solar cycles hardly impact month-to-month surface temperature. The third predictor included here is the multivariate index for El Niño Southern Oscillation, a circulatory phenomenon that affects ocean-atmosphere heat transfer. Including this predictor explains more of the variance in monthly temperature but cannot account for its multidecade upward trend. After including atmospheric CO_2 as a fourth predictor, however, we see substantial improvement. These four indicators together yield a surprisingly good fit (adjusted $R^2 = 0.73$) to the jagged sequence of 371 monthly temperatures.

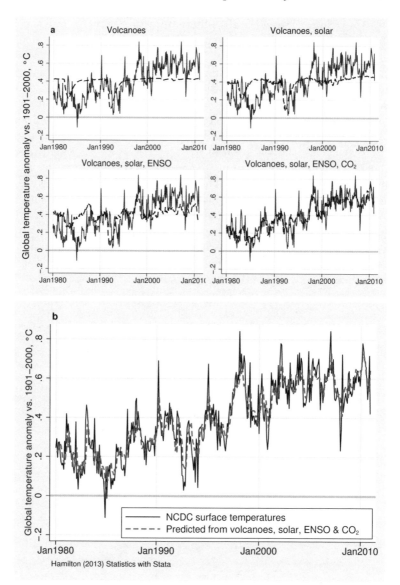

Figure 5.1. a) Global surface temperature regressed on one, two, three, and four lagged predictors; and b) a time-series regression with all four predictors plus ARIMA(1,0,1) disturbances.

Unsurprisingly, residuals from the four regressions in figure 5.1a exhibit significant autocorrelation. Figure 5.1b depicts a statistically better autoregressive integrated moving-average (ARIMA) model that uses the same four lagged predictors plus ARIMA(1,0,1) disturbances, resulting in white-noise residuals. The *Statistics with Stata* examples graphed in figure 5.1 present a highly simplified version of the major climate discovery that recent warming cannot be explained by natural factors without taking greenhouse gasses into account (IPCC [2013]; for a statistical analysis that inspired my *Statistics with Stata* example, see Foster and Rahmstorf [2011]). At the same time, this example makes the important point that greenhouse gasses are not the only factor. El Niño Southern Oscillation in particular influences air temperature variations and short-term trends, such as the slowdown observed in surface warming during the 2000s when more heat went into the deep oceans (England et al. 2014).

Along with such autoregressive moving average with exogenous inputs (ARMAX) time series, Stata's growing capabilities for multilevel or mixed-effects modeling, recently extended to generalized structural equation modeling (GSEM), make it well suited for interdisciplinary research that combines social and physical data. Such integrated research has been widely advocated for studies of socioenvironmental systems, but rigorous integration faces many complications in practice (Huntington et al. 2007). Stata's new statistical methods help on this frontier.

Figure 5.2 illustrates two integrated studies, too recent to be in *Statistics with Stata*, that combine survey research with climatology. Telephone survey responses are merged, by date of interview, with climate data from weather station records. Figure 5.2a is a charismatic `marginsplot` image often called the "Zorro graph". Initially published in the journal article "Blowin' in the wind: Short-term weather and belief in anthropogenic climate change" (Hamilton and Stampone 2013), this graph drew blog and media attention and was reproduced in *Science* (2013). Our data began with more than 5,000 random-sample interviews of New Hampshire residents, conducted on 99 days over 2010 to 2012. The survey questions asked what respondents personally believe about climate change. Is it happening now and caused mainly by human activities, or is it happening now and caused mainly by natural forces? Is it not happening now? Or do you not know? The y axis in figure 5.2a is the adjusted marginal probability of the response being that it is happening now and is caused by humans.

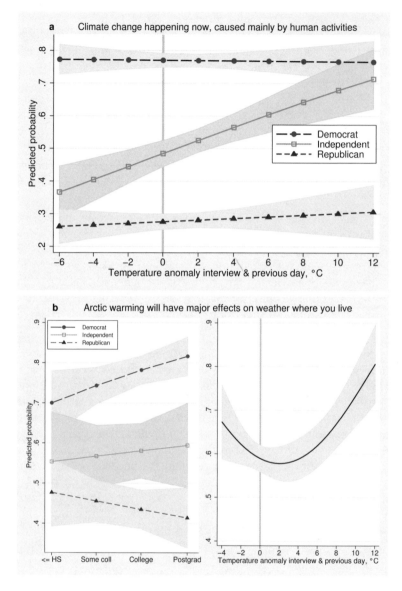

Figure 5.2. a) Probability of now/human response as function of temperature and party, adjusted for other variables (Hamilton and Stampone 2013). b) Probability of Arctic warming affects weather response as function of education and party, or of temperature, adjusted for other variables (Hamilton and Lemcke-Stampone 2014).

My colleague (New Hampshire State Climatologist, Mary Stampone) and I constructed a statewide temperature anomaly index by averaging records from five geographically diverse weather stations. Despite geographical variations in absolute temperatures, temperature anomalies (daily deviation from the 1981–2010 average for each station and date) correlate well across stations, so the first principal component explains 84% of their variance over the study period. The 2-day mean temperature anomaly defines the x axis in 5.2a.

Do unseasonably warm or cool temperatures affect general climate change beliefs? It seems they do, even if we control for other factors (in a mixed-effects logit regression, `melogit`) and include gender, age, education, political party, and season as predictors, with random intercepts by survey. But these temperature effects, which are strongest for just a two-day window, mostly impress Independent voters. Democrats tend to believe in anthropogenic climate change no matter what today's temperature. Republicans tend not to believe it, regardless of temperature. These groups form the top and bottom curves in 5.2a. Unaffiliated or Independent voters, on the other hand, are "blowing in the wind"—they believe in anthropogenic climate change on unseasonably warm days, but they do not believe it on cool ones. That striking result suggests a climate-change counterpart to polling studies that have shown volatility among uncommitted (frequently low-information) voters, who can swing election outcomes at the last minute.

Figure 5.2b shows results from subsequent New Hampshire surveys that asked a different question: if the Arctic region becomes warmer in the future, do you think that will have major effects, minor effects, or no effects on the weather where you live? (Hamilton and Lemcke-Stampone 2014). The rightmost panel in 5.2b again depicts a temperature-anomaly effect, this time quadratic: people are more likely to believe an Arctic and weather connection when interviewed on unseasonably warm or cool days. The curvilinear relationship unscientifically mirrors media discussion of scientific studies linking Arctic warming to mid-latitude weather extremes (for example, Francis and Vavrus [2012]; Screen and Simmonds [2014]).

The left-hand panel in figure 5.2b shows an `education` × `party` interaction from the same logit model. Belief in an Arctic and weather connection increases with education among Democrats, but decreases with education among Republicans. Similar interactions have been observed across other survey datasets; they particularly affect climate-belief questions. Stata-powered examples of such interactions include one- or two-survey analyses (Hamilton 2008, 2011, 2012; Hamilton, Cutler, and Schaefer 2012) along with others that use mixed-effects modeling to test environmental effects in data pooled from many surveys (Hamilton and Keim 2009; Hamilton, Colocousis, and Duncan 2010; Hamilton et al. 2014; Hamilton and Safford 2015). Interactions have long been a staple of social science research, but Stata's factorial notation for regression predictors, such as `c.education##ib2.party` (interactions and main effects with `education` treated as a continuous variable, `party` as a set of indicator or dummy variables, and `party = 2` as the base or reference category), makes it much easier to test such hypotheses. At the same time, `marginsplot` sets what should become a new standard for graphically displaying interactions with their confidence bands.

5.4   Future editions

As the publishing industry moves rapidly sideways, it is hard to forecast the direction of *Statistics with Stata*. Future editions could incorporate more integrated research, along with basic tricks and selected new features with broad applications. There should also be more examples showing how Stata's graphical capabilities can help craft clear and accurate yet media-friendly presentations of results. Public dissemination has experienced growing recognition as an important dimension of research. Figure 5.3 shows a simple but evocative `hbar` graph (originally from Hamilton [2014]) that has traveled widely, drawing appearances or mention as far afield as *Mother Jones* (Mooney 2014), *Popular Science* (Diep 2014), and *The Guardian* (Harman 2014).

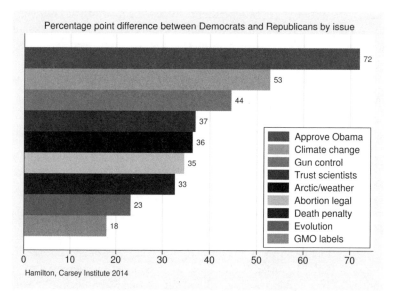

Figure 5.3. Partisan differences on issues in a New Hampshire statewide poll (Hamilton 2014)

Other graphics that I also created using Stata have been copied on blogs, and one made an unacknowledged appearance in testimony to the British Parliament. After that, I briefly made a practice of noting my name and the reference within graphs I published, but I soon discovered that some people would just remove it. Following daily developments as a new research article enters the spin cycle of instant media, blog, and Internet arguments has become a popular topic in my classes. Using search engines to track the spread of true and false memes, or the spread of graphs out of context, can ruefully turn this spin cycle into data as well. Illustrations of such postresearch odysseys might show up in a future *Statistics with Stata* as further examples of data that have stories to tell.

References

Diep, F. 2014. When scientific facts get divided down party lines. http://www.popsci.com/article/science/when-scientific-facts-get-divided-down-party-lines.

England, M. H., S. McGregor, P. Spence, G. A. Meehl, A. Timmermann, W. Cai, A. S. Gupta, M. J. McPhaden, A. Purich, and A. Santoso. 2014. Recent intensification of wind-driven circulation in the Pacific and the ongoing warming hiatus. *Nature Climate Change* 4: 222–227.

Foster, G., and S. Rahmstorf. 2011. Global temperature evolution 1979–2010. *Environmental Research Letters* 6: 044022.

Francis, J. A., and S. J. Vavrus. 2012. Evidence linking Arctic amplification to extreme weather in mid-latitudes. *Geophysical Research Letters* 39: L06801.

Hamilton, L. 2005. A short history of Statistics with Stata. *Stata Journal* 5: 35–37.

Hamilton, L. C. 1990a. *Modern Data Analysis: A First Course in Applied Statistics.* Pacific Grove, CA: Brooks/Cole.

———. 1990b. *Statistics with Stata.* Belmont, CA: Brooks/Cole.

———. 1992. *Regression with Graphics: A Second Course in Applied Statistics.* Pacific Grove, CA: Brooks/Cole.

———. 1993. *Statistics with Stata (Updated for Version 3).* Belmont, CA: Brooks/Cole.

———. 1998. *Statistics with Stata (Updated for Version 5).* Belmont, CA: Brooks/Cole.

———. 2002. *Statistics with Stata (Updated for Version 7).* Pacific Grove, CA: Duxbury.

———. 2004. *Statistics with Stata (Updated for Version 8).* Belmont, CA: Brooks/Cole.

———. 2008. Who Cares about Polar Regions? Results from a Survey of U. S. Public Opinion. *Arctic, Antarctic, and Alpine Research* 40: 671–678.

———. 2009. *Statistics with Stata (Updated for Version 10).* Belmont, CA: Brooks/Cole.

———. 2011. Education, politics and opinions about climate change evidence for interaction effects. *Climatic Change* 104: 231–242.

———. 2012. Did the Arctic ice recover? Demographics of true and false climate facts. *Weather, Climate, and Society* 4: 236–249.

———. 2013. *Statistics with Stata (Updated for Version 12).* Belmont, CA: Brooks/Cole.

———. 2014. Do you trust scientists about the environment? News media sources and politics affect New Hampshire resident views. Regional Issue Brief No. 40. Durham, NH: Carsey Institute, University of New Hampshire. http://scholars.unh.edu/cgi/viewcontent.cgi?article=1213&context=carsey.

Hamilton, L. C., C. R. Colocousis, and C. M. Duncan. 2010. Place effects on environmental views. *Rural Sociology* 75: 326–347.

Hamilton, L. C., M. J. Cutler, and A. Schaefer. 2012. Public knowledge and concern about polar-region warming. *Polar Geography* 35: 155–168.

Hamilton, L. C., J. Hartter, T. G. Safford, and F. R. Stevens. 2014. Rural environmental concern: Effects of position, partisanship, and place. *Rural Sociology* 79: 257–281.

Hamilton, L. C., and B. D. Keim. 2009. Regional variation in perceptions about climate change. *International Journal of Climatology* 29: 2348–2352.

Hamilton, L. C., and M. Lemcke-Stampone. 2014. Arctic warming and your weather: public belief in the connection. *International Journal of Climatology* 34: 1723–1728.

Hamilton, L. C., and T. G. Safford. 2015. Environmental views from the coast: Public concern about local to global marine issues. *Society and Natural Resources* 28: 57–74.

Hamilton, L. C., and C. L. Seyfrit. 1993a. sqv8: Interpreting multinomial logistic regression. *Stata Technical Bulletin* 13: 24–28. Reprinted in *Stata Technical Bulletin Reprints*, vol. 3, pp. 176–181. College Station, TX: Stata Press.

———. 1993b. Town-village contrasts in Alaska youth aspirations. *Arctic* 46: 255–263.

Hamilton, L. C., and M. D. Stampone. 2013. Blowin' in the wind: Short-term weather and belief in anthropogenic climate change. *Weather, Climate, and Society* 5: 112–119.

Harman, G. 2014. Climate change: How businesses can deal with America's most divisive issue. http://www.theguardian.com/sustainable-business/2014/jun/12/climate-change-global-warming-businesses-divisive-issue-america.

Huntington, H. P., L. C. Hamilton, C. Nicolson, R. Brunner, A. Lynch, A. E. J. Ogilvie, and A. Voinov. 2007. Toward understanding the human dimensions of the rapidly changing arctic system: Insights and approaches from five HARC projects. *Regional Environmental Change* 7: 173–186.

IPCC. 2013. *Climate Change 2013—The Physical Science Basis. Summary for Policy Makers*. Geneva, Switzerland: Intergovernmental Panel on Climate Change.

Mooney, C. 2014. Poll: Tea Party members really, really don't trust scientists. http://www.motherjones.com/environment/2014/05/tea-party-climate-trust-science.

Science. 2013. Independents follow where the wind blows. *Science* 339: 496.

Screen, J. A., and I. Simmonds. 2014. Amplified mid-latitude planetary waves favour particular regional weather extremes. *Nature Climate Change* 4: 704–709.

5.5   About the author

Lawrence Hamilton is a professor of sociology at the University of New Hampshire, and he is a senior fellow at the Carsey School of Public Policy. He wrote the first *Statistics with Stata* during the MS-DOS era of desktop computers, and has since had seven newer editions published. Hamilton has conducted interdisciplinary research around the circumpolar North, with case studies or comparative overviews of resource-dependent communities in Alaska, Newfoundland, Greenland, Iceland, the Faroe Islands, and Norway. He is an active participant in national and international working groups on the human dimensions of Arctic change. As a Carsey School fellow, Hamilton works on the design and analysis of surveys such as the Community and Environment in Rural America, Communities and Forests in Oregon, and the quarterly Granite State Poll projects. The surveys track public knowledge and perceptions about science, resources, and climate.

6 Stata's contribution to epidemiology

Stephen J. W. Evans
London School of Hygiene and Tropical Medicine
London, UK

Patrick Royston
MRC Clinical Trials Unit
University College London
London, UK

6.1 Introduction

At a meeting of the American Public Health Association in Washington, DC, on November 18, 1985, one of us (Stephen J. W. Evans) presented an article on "Trends in birth weight distribution: 1970–83".

The data for that article were analyzed on a mainframe computer using SAS, and the programs and data were prepared on 80-column punched cards. The main statistical analysis programs were being rewritten around that year to run on the new personal computers. Stephen J. W. Evans had just recently obtained a Hyperion microcomputer, which was "luggable"—it weighed over 8kg, had two 5.25-inch floppy disks, and was largely used for word processing. Luggable PC-compatible computers at that time did not have "hard" disks.

At the 1985 meeting in Washington, DC, Bill Gould had a stand where he demonstrated statistical software running on a twin floppy IBM PC (or possibly a Compaq luggable). This showed extremely fast data analysis, and the use of a scatterplot matrix plotted at high speed was extremely eye-catching! The Stata program was on one 5.25-inch floppy disk and the data were on another. Two demonstration disks were available (and retained for posterity—see figure 6.1), and they seemed to be based on Stata 1.3 in that they used "real" (as opposed to character-based) graphics. These demonstration disks were shown to Doug Altman, who may have been the first to actually purchase Stata for medical research in the UK (if not Europe). He subsequently showed it to Patrick Royston, and the rest is history!

Figure 6.1. Stata floppy disk

6.2   Epidemiology and Stata

Nicholas Cox (2005) provided a clear history of Stata's first 20 years. There are many
features of Stata that are very useful to all users and not specifically for epidemiology,
and these have been covered well by others.

The usual statistical analyses were possible on the demonstration disks, and the
speed advantage was at least partly because the disks' entire dataset was kept in RAM.
This was, at that time, both an obvious benefit and an obvious problem. Epidemiology
often used datasets that exceeded the memory capacity of (Microsoft) DOS and, hence,
of Stata.

However, datasets, even large ones, were often summarized in multiway tables. Survival data, especially with large numbers included, were often summarized in what one

might describe as "clinical life tables", giving data in year intervals like an actuarial life table. Indeed, while Stata provided Kaplan–Meier survival analysis with great graphics from the early days of Stata 2.0 (1988), the first edition of the *Stata Technical Bulletin* in 1991 included `lftbl`, a program to implement life-table analysis with a form of log-rank test. It departed from Stata standards in that it was driven by interactive responses when setting variables and parameters. As far as we recall, while the output documented the choices made, repeating an analysis required retyping the responses. The second *Stata Technical Bulletin* illustrated a general feature of Stata with the arrival of `ltable`, which was a Stata-supported life-table program and was command driven. It showed that users who write really useful programs can find their programs taken by Stata and improved, particularly in making them robust to errors, more accurate, faster, and generally more efficient. "Any fool can build a bridge that will stand up, it takes an engineer to build a bridge that will just stand up"; "Any fool can write a program that works, it takes a real programmer to write one that doesn't fall over when it's given errors in the input".

Ken Rothman provided instructions for key epidemiological analyses using a programmable hand calculator in 1979. A new edition of the short book (with Boise) appeared in 1984, updated for the HP-41CV calculator. It was not until 1990 that the first issue of the journal *Epidemiology* appeared. This journal, edited by Rothman, reflected the increasing demand for a journal that would "foster critical discussion of principles, methods, results, and implications of epidemiologic research".

Stata's arrival in this environment had immediate and lasting appeal for epidemiologists and statisticians. Others have discussed general reasons for the popularity of Stata, notably the responsiveness of StataCorp to user requests and the ability to develop programs using Stata commands in an efficient and transparent manner.

For epidemiologists, the impact of the immediate commands associated with `epitab` has been extensive. These commands were available from Stata 3.0 in 1992. The ability to take data from tables, published or unpublished, and provide an analysis estimating odds or risk and incidence-rate ratios was so useful. It was the sort of usage of Stata that would not generally be cited in a publication, but made the day-to-day work of the epidemiologist easier. Similarly, the commands to calculate confidence intervals for proportions were a big help. Before then, many used the famous Geigy Scientific tables, which were a useful source of statistical table. While, particularly in the early days, Stata was probably not as sophisticated in its data manipulation facilities as SAS, it enabled a user to get much closer to the data. Tukey's exploratory data analysis paradigm was more used by medical statisticians working with clinical or laboratory data, but the design of Stata encouraged that kind of familiarity with the raw data for researchers actually doing epidemiological analysis. Stem-and-leaf plots were available very early, and Stata was great for checking data, especially with its easy use of scatter plots.

The command `logit` was introduced in Stata 1.5 in 1987. Logistic regression, and the introduction of clear commands for doing conditional logistic regression, were heavily used at this time and the latter also appeared in Stata 3.0. In the early days, many people used programs that used Cox regression for conditional logistic regression. For many years after the classic 1972 article by Cox, Cox regression was done using FOR-TRAN programs. One of these programs was developed by Richard Peto, and the code was widely available. Stephen J. W. Evans added "line printer" graphics to that program to include estimated survival curves from a Cox regression in his MSc project in 1977–1978. The first edition of Kalbfleisch and Prentice's book *The Statistical Analysis of Failure Time Data* (1980) included FORTRAN code to do Cox regression (unfortunately with transposition of some sections of the code, which frustrated my early attempts to make it work). In the same year, Breslow and Day's classic book *Statistical Methods in Cancer Research: Volume I—The Analysis of Case–Control Studies* (1980) included FORTRAN code to perform the relevant analyses including logistic regression. Using these FORTRAN programs and compiling the FORTRAN code was cumbersome, and although FORTRAN compilers were available for PCs under DOS, use of these by epidemiologists usually required a statistician with good programming skills. Stata (and other statistical package programs) changed all that, and being able to do Cox regression with time-updated or time-varying analyses with results that could be trusted was a major advance. Stata was very responsive to ensure that survival analysis was done well, was easy to specify, and produced output that was comprehensible and reasonably comprehensive.

In the late 1980s and early 1990s, published articles provided statistical methods, but they did not always include the software used to perform the analysis. In the UK, GLIM was a widely used package that provided sophisticated analyses. GLIM was not very user friendly (it was sometimes accused of encouraging incomprehensible "write-only code"), which was sometimes an advantage because clever statisticians were usually required to do the analysis! This provided employment for statisticians. I recall at an open-discussion meeting where analysis of different datasets was discussed, David Cox asked a presenter what method was used to do his or her analysis. The presenter replied, "I used GLIM", and Cox gently asked, "but what exact statistical method did you use?" Once again, the presenter responded "I used GLIM". The gracious David Cox did not pursue it further! Stata enabled the application of new and (for the time) sophisticated methods, and it was easy to show exactly what statistical method had been used.

The lucid textbook by Doug Altman (1990), *Practical Statistics for Medical Research*, may have helped increase Stata's popularity among those working in epidemiology. Although the book was targeted more at clinical research, it included short sections on basic epidemiology designs and illustrated most of the analyses and data using Stata graphics.

It was probably Stata 3.0 (1992) and Stata 3.1 (1993) that had the greatest impact on epidemiology, and practicing epidemiologists began to actually use Stata. As noted above, the commands `logistic` and `clogit`, together with those included in `epitab`, appeared in version 3.0. The three manuals for version 3.1 can still be found on the shelves of leading epidemiologists today!

Use of sophisticated methods has grown in epidemiology. Propensity scores, which have been used increasingly, with graphical display of results are easy, so good practitioners use these for model checking.

For all Stata users, perhaps especially those dealing with large datasets (as in much of epidemiology), the advent of cheaper main memory and software in recent years that allowed access to large amounts of RAM has been very helpful. Now 64-bit operating systems allow enormous datasets to be included in main memory. The use of virtual memory may make processing possible where previously it was impossible.

Stata has introduced two particular features that make it practical and appealing to use in epidemiology, specifically in a regulated environment. The first feature, which may be unique in standard statistical software, is the capability to use ICD-9-CM codes, which were introduced into Stata in 2000. Epidemiology, especially pharmacoepidemiology, has made major use of United States and other countries routinely collected health records. Many of these records have used ICD-9 codes to define and classify medical diagnoses. The commands offered for this by Stata are helpful, especially in checking data. In all statistical and epidemiological analyses, checking the data is a vital, time-consuming, and sometimes tedious task. However, it is usually better to have a suboptimal analysis of correct data than an optimal analysis of incorrect data. In epidemiology, these tasks can be a major aspect of the computing work, because the volume of data may be extremely large.

The second feature facilitates making datasets transferable between SAS and Stata via the SAS XPORT file format. The Stata commands `export sasxport` and `import sasxport` are obviously of general use, but the U.S. Food and Drug Administration requires that datasets supplied to them are in SAS format. This has led to the mythology that the only software acceptable to the U.S. Food and Drug Administration is SAS, which is simply not true. Regulatory authorities are increasingly using epidemiology, especially in "risk management plans" being used around the world to monitor the safety of medicines and vaccines. These commands offered by Stata help regulatory authorities ensure that products meet requirements. However, the pharmaceutical industry tends to use SAS because it is a "safe" option, rather than because of inherent advantages.

A major feature in epidemiology in the last 30 years has been the rise in meta-analysis. It could easily be argued that meta-analysis of randomized clinical trials is part of epidemiology, and most text books of epidemiology have important sections discussing it. It is one area where Stata does not (yet) have official commands, but they have extensive support to users writing such commands. They have gone as far as to publish a book through Stata Press—*Meta-Analysis in Stata: An Updated Collection from the Stata Journal*, edited by Jonathan Sterne (2009). Stata competes with software that is solely for meta-analysis, and it has been heavily used by epidemiologists.

6.3   The future

The London School of Hygiene and Tropical Medicine has one of the largest, if not the largest, groups of medical statisticians and epidemiologists in Europe. Stata has been used for teaching and for research for over 20 years and tends to be the standard software used by most practitioners. Use of "R" is increasing among statisticians. Its noncommercial nature is financially appealing but, correspondingly, the quality (certification) of the software, an enduring feature of Stata, may not always be reliable.

Propensity-score methods are used frequently and will continue to be used. Stata provides facilities for using these methods easily within the current official release. The relatively new commands using structural equation modeling (`sem`) and treatment effects through `teffects` are at the cutting edge of epidemiological methods. Causal effects are always desired in epidemiology, and mediation methods are a new fashion and likely to increase in use. There will likely be further developments in this area. When the methods have stabilized, or have at least matured to the point of having generally agreed approaches, they will become part of official Stata.

The *Stata Journal* continues to be a source of the latest methods. It includes articles with clear descriptions on how to use official Stata commands in sensible ways or with discussions on new user-written commands, such as the article by Daniel, De Stavola, and Cousens (2011) introducing `gformula`, a tool using the g-computation formula to estimate causal effects in the presence of time-varying confounding or mediation.

The future is bright for epidemiologists using Stata. In the earliest days, Stata was catching up with the methods in use. During most of the 90s, Stata and epidemiological methods seemed to be advancing together, but it seems now that official Stata and new methods available in repositories or published in the *Stata Journal* are ahead of most epidemiological practice. Practice will catch up with current methods, but new methods will continue to appear. Some of these new methods will appear first in "R", but Stata seems to be very quick to provide more user-friendly software with very little delay. The complexity of the methods has risen dramatically, and the need for expository articles and textbooks will increase.

It will be interesting to see if Stata adds official meta-analysis commands. If so, will they find a good way of doing meta-analysis of observational (as opposed to randomized) data that allows for the greater uncertainty beyond sampling variation in summary estimates derived from nonrandomized or even mixed (randomized and nonrandomized) data? Demand for certainty should not allow just the use of large or "big data" samples to introduce a spurious certainty.

References

Altman, D. G. 1990. *Practical Statistics for Medical Research*. London: Chapman & Hall.

Breslow, N. E., and N. E. Day. 1980. *Statistical Methods in Cancer Research: Volume I—The Analysis of Case–Control Studies*. Lyon: International Agency for Research on Cancer.

Cox, D. R. 1972. Regression models and life-tables. *Journal of the Royal Statistical Society, Series B* 34: 187–220.

Cox, N. J. 2005. A brief history of Stata on its 20th anniversary. *Stata Journal* 5: 2–18.

Daniel, R. M., B. L. De Stavola, and S. N. Cousens. 2011. gformula: Estimating causal effects in the presence of time-varying confounding or mediation using the g-computation formula. *Stata Journal* 11: 479–517.

Evans, S. J. W. 1985. Trends in birth weight distribution: 1970–83. http://www.cdc.gov/nchs/data/misc/ice2_85_2.pdf.

Kalbfleisch, J. D., and R. L. Prentice. 1980. *The Statistical Analysis of Failure Time Data*. New York: Wiley.

Krakauer, H., and J. Stewart. 1991. ssa1: Actuarial or life-table analysis of time-to-event data. *Stata Technical Bulletin* 1: 23–25. Reprinted in *Stata Technical Bulletin Reprints*, vol. 1, pp. 200–202. College Station, TX: Stata Press.

Rothman, K. J., and J. D. Boice. 1979. *Epidemiologic Analysis With a Programmable Calculator*. Bethesda, MD: U.S. Dept. of Health, Education, and Welfare, Public Health Service, National Institutes of Health.

Rothman, K. J., J. D. Boice, and H. Austin. 1984. *Epidemiologic Analysis With a Programmable Calculator: With an Appendix for the HP-41CV by H. Austin*. Boston, MA: Epidemiology Resources, Inc.

Sterne, J. A. C., ed. 2009. *Meta-Analysis in Stata: An Updated Collection from the Stata Journal*. College Station, TX: Stata Press.

6.4 About the authors

Stephen Evans is a medical statistician, and currently he is Professor of Pharmacoepidemiology at the London School of Hygiene and Tropical Medicine. He has extensive expertise in the development, implementation, and teaching of appropriate methods in pharmacoepidemiology and pharmacovigilance. He was president of the International Society of Pharmacoepidemiology in 2010–2011.

After training in physics and chemistry he worked in statistics and computing at The London Hospital and Medical College, then Head of Epidemiology at the UK Medicines

Control Agency, where he dealt with several major drug safety issues, such as hormone replacement therapy and breast cancer; vitamin K and childhood cancer; measles–mumps–rubella and autism. He also developed some statistically-based methods for examining regulatory databases of spontaneous reports.

He is currently a coopted Expert member of the Pharmacovigilance Risk Assessment Committee Party at the European Medicines Agency and was a member of the WHO Global Advisory Committee on Vaccine Safety until June 2012. He was a statistical adviser to the British Medical Journal for 20 years. He has been on various editorial boards, including the British Journal of Clinical Pharmacology and is an Associate Editor of Pharmacoepidemiology and Drug Safety.

Professor Royston's interests center around statistical modeling and its medical applications, modeling continuous predictors, survival analysis, methodology of clinical trial design, and imputation of missing covariate data. He has a strong interest in developing and disseminating software implementation of new research methods in the form of Stata programs. He has co-authored two books; *Multivariable Model-Building. A pragmatic approach to regression analysis based on fractional polynomials for modeling continuous variables* (2008) with Willi Sauerbrei of Freiburg University, Germany and *Flexible Parametric Survival Analysis Using Stata: Beyond the Cox Model* (2011) with Paul Lambert of Leicester University, UK.

Professor Royston is an honorary professor of statistics in the Department of Statistical Science at University College London and a senior statistician at the MRC Clinical Trials Unit in London.

7 The evolution of veterinary epidemiology

Ian R. Dohoo
University of Prince Edward Island
Charlottetown, Canada

7.1 History of veterinary epidemiology

Veterinary epidemiology has a long and proud history of contributions to both animal and human health. One specific example is Daniel Salmon (United States Bureau of Animal Industry founder and after whom *Salmonella* was named) and Frederick Kilborne's application of epidemiologic methods when studying "Texas Fever" in the 1880s. After studying the similar geographical distributions of the disease and a cattle tick (*Boophilus annulatus*) and completing some controlled field trials, they determined that the disease was spread by this insect vector, although the direct causal agent of the disease—a parasite, *Babesia bigemina*—was not discovered until many years later (Schwabe 1984). This was the first demonstration of a parasite requiring development within a vector before transmission.

More recently, veterinary epidemiologist have played a lead role in determining that bovine spongiform encephalopathy is transmitted by feeding inadequately rendered livestock offal to cattle (Wilesmith et al. 1988) and in globally eradicating Rinderpest (a highly fatal and contagious viral disease in cattle) in 2010.

Much of the early epidemiologic work focused on determining the natural history of infectious diseases. In the 1970s, increased emphasis was placed on studying production-related diseases, with the "Bruce County Project" (Martin et al. 1980; Martin 1979) and the "Benchmark Project" (McDermott et al. 1991) being two notable Canadian examples in the beef industry. The "Bruce County Project" was one of the first studies to clearly establish the multifactorial nature of respiratory disease in beef cattle and the potentially deleterious effects of some prophylactic procedures used in feedlot steers. These projects placed a new emphasis on large-scale data-collection efforts over multiple years. More recently, the "National Cohort of Dairy Farms" project (Reyher et al. 2011) represents one of the largest multiyear data-collection efforts in any dairy industry.

7.2   Development of quantitative veterinary epidemiology

In 1966, Dr. Calvin Schwabe (widely known as the "father of modern veterinary epidemiology") joined the University of California and established the Department of Epidemiology and Preventive Medicine. His textbook *Epidemiology in Veterinary Practice* (1977) was the first text to take a distinctly quantitative approach to veterinary epidemiology. Graduate students of Dr. Schwabe spread this new approach across North America and to Europe. Most notably, Dr. Wayne Martin established the Department of Population Medicine at the University of Guelph and was the lead author on the first textbook to focus on quantitative methods in veterinary epidemiology—*Veterinary Epidemiology, Principles and Methods* (Martin, Meek, and Willeberg 1987). Strong programs in quantitative veterinary epidemiology were soon established in many other schools, including Cornell University (United States), Copenhagen University (Denmark), Massey University (New Zealand), Utrecht University (The Netherlands), and the University of Prince Edward Island (Canada). A unique feature of these programs has been the effective integration of statisticians into the epidemiology faculty.

More recently, other texts in the discipline have contributed to the growth of the discipline: *Veterinary Epidemiology* (Thrusfield 2007), *Veterinary Clinical Epidemiology* (Smith 2005), and *Veterinary Epidemiologic Research* (Dohoo, Martin, and Stryhn 2009b). This last book emphasized and promoted quantitative veterinary epidemiology methods and used Stata for nearly all of the examples presented in the text.

7.3   Distinctive aspects of veterinary epidemiology

The development of quantitative veterinary epidemiologic methods has led to a significant expansion of the discipline into non-infectious diseases. The majority of the published research in the discipline deals with health and productivity of production animals (cattle, pigs, poultry, fish, etc.) and horses, rather than companion animals (dogs, cats, etc.), but this is a function of the limited availability of funding for companion-animal research rather than a comment on the applicability of the methods.

In general, the methods used in veterinary epidemiology are similar or identical to those used in medical ("human") epidemiology. The two branches of the discipline are getting even closer together with the development of the "one-health" or "one-medicine" concept. Once again, Dr. Calvin Schwabe played a pivotal role with the publication of his classic text *Veterinary Medicine and Human Health* (Schwabe 1984). Since then, the movement has continued to expand with an independent organization dedicated to its development—the One Health Initiative (http://www.onehealthinitiative.com). The similarity of the methods is evidenced by the fact that the text *Veterinary Epidemiologic Research* (Dohoo, Martin, and Stryhn 2009b) was recently converted into a textbook for human-health practitioners—*Methods in Epidemiologic Research* (Dohoo, Martin, and Stryhn 2009a)—with the conversion of the animal-health examples to human-health examples.

Nonetheless, there are many areas of importance to veterinary epidemiology for which Stata provides an excellent set of tools. These include:

- analysis of clustered (multilevel) data
- methods used in surveillance
- survival-analysis methods
- meta-analysis (used in conjunction with risk analysis)

Virtually all animal-health data (particularly those relating to production animals) are clustered and often have repeated measures. For example, data may be collected from multiple lactations (repeated measures) from dairy cows that are clustered within herds (hierarchical data). Handling the multilevel structure of these data correctly is crucial to veterinary epidemiology. Early work by McDermott (McDermott, Schukken, and Shoukri 1994), followed by efforts by Gröhn (Gröhn et al. 1999) and Schukken (Schukken et al. 2003), highlighted the importance of this issue. More recently, analysis of dairy-production data with four-levels (lactation, cow, herd, region) from Reunion Island was used to explore possible approaches to these analyses (Dohoo et al. 2001). An evaluation of the current state-of-the-art methods for dealing with clustered data was recently published (Stryhn and Christensen 2014). Relative to other major software packages, Stata was rather slow with the implementation of methods for dealing with clustered continuous data, so the addition of `xtmixed` in Stata 9 (now `mixed`) was a very welcome addition. By contrast, Stata has long been a leader in the implementation of methods for analyzing clustered discrete data, with the `gllamm` command contributed by Rabe-Hesketh and coworkers being, in my opinion, the "gold standard" of methods for these types of analyses. Stata routines such as `xtlogit` (introduced in Stata 6), `xtmelogit` (introduced in Stata 10), and `meglm` (introduced in Stata 13) greatly simplify and facilitate the analysis of clustered discrete data. Another unique feature of Stata is the widespread availability of cluster–robust standard errors (option `vce(cluster)`) across estimation procedures to allow for adjustment of standard errors to account for clustering.

With the globalization of agricultural production and marketing, a large proportion of effort in regulatory veterinary medicine is directed toward surveillance for animal diseases. Methods for the analysis of such survey data are critical to the discipline (Dargatz and Hill 1996), and the set of survey analysis commands (`svy` commands) has been a very useful contribution in this area.

Although the "survival" of meat animals is often dictated by the production cycle (for example, beef animals are usually slaughtered at between 1 and 2 years of age), the longevity of animals in many production settings (for example, longevity of dairy cows) is key to the profitability of the enterprise. This, combined with the fact that time to onset of disease conditions within a production cycle (for example, time to onset of clinical mastitis within a dairy cow's lactation [Elghafghuf et al. 2014]) has resulted in a substantial rise in interest in survival analysis methods in veterinary epidemiology. The `st` set of commands in Stata is one of the most complete and readily accessible sets of survival analysis commands. The recent addition of `stmixed` (2014) to perform flexible parametric modeling of survival data (Crowther, Look, and Riley 2014) while accounting for the clustered nature of the data is likely to be a major contribution to veterinary epidemiology.

Historically, meta-analysis has not played as important a role in veterinary medicine as it has in human medicine. This has largely been due to the fact that veterinary medicine has a much smaller body of literature available and, therefore, there has been less need for methods for summarizing results across multiple studies. However, that is changing. As with the need for surveillance to support regulation of movement of animals and animal products, there has been a dramatic rise in the requirement for formal risk assessments to underpin decisions for such product movements (see Understanding the World Trade Organization Agreement on Sanitary and Phytosanitary Measures—http://www.wto.org/english/tratop_e/sps_e/spsund_e.htm). Although Stata is not used for risk analyses, meta-analysis is increasingly required to provide data to support these risk analyses and other regulatory decisions. User contributed routines for conducting meta-analyses with Stata (for example, `metan`), combined with Stata's flexible data-handling procedures, make this an ideal program for these activities. The first animal-health related meta-analysis done in North America used Stata to evaluate the effects of recombinant bovine somatotropin (rBST or "bovine growth hormone") on dairy-cattle health (Dohoo et al. 2003a,b). This review led the Canadian government to decline the request for registration of the product in Canada.

7.4   Use of Stata in veterinary epidemiology

Preventive Veterinary Medicine is the primary journal for publication of veterinary epidemiologic research. A review of papers published in 1993 (when Stata was 8 years old), 2003, and 2013 shows the growth in the use of Stata within veterinary epidemiology. The proportion of articles citing Stata use in those three years was 0%, 13%, and 24%, respectively.

Stata has been increasingly accepted in graduate programs in veterinary epidemiology. Its strengths in epidemiologic analyses (particularly `epitab`) combined with its strong suites of programs for fitting multivariable models and for dealing with clustered data have made it a popular choice among veterinary epidemiologists. (On a personal note—it was the University of Prince Edward Island's desire to move from using three separate programs to one program in the graduate program that motivated me to do a review [in the early 1990s] of all major statistical programs and to, as a result, identify Stata as a program with good epidemiologic and statistical capabilities.)

References

Crowther, M. J., M. P. Look, and R. D. Riley. 2014. Multilevel mixed effects parametric survival models using adaptive Gauss–Hermite quadrature with application to recurrent events and individual participant data meta-analysis. *Statistics in Medicine* 33: 3844–3858.

Dargatz, D. A., and G. W. Hill. 1996. Analysis of survey data. *Preventive Veterinary Medicine* 28: 225–237.

Dohoo, I., W. Martin, and H. Stryhn. 2009a. *Methods in Epidemiologic Research.* Charlottetown, Canada: VER Inc.

———. 2009b. *Veterinary Epidemiologic Research.* 2nd ed. Charlottetown, Canada: VER Inc.

Dohoo, I. R., L. DesCôteaux, K. Leslie, A. Fredeen, W. Shewfelt, A. Preston, and P. Dowling. 2003a. A meta-analysis review of the effects of recombinant bovine somatotropin. 2. Effects on animal health, reproductive performance, and culling. *Canadian Journal of Veterinary Research* 67: 252–264.

Dohoo, I. R., K. Leslie, L. DesCôteaux, A. Fredeen, P. Dowling, A. Preston, and W. Shewfelt. 2003b. A meta-analysis review of the effects of recombinant bovine somatotropin. 1. Methodology and effects on production. *Canadian Journal of Veterinary Research* 67: 241–251.

Dohoo, I. R., E. Tillard, H. Stryhn, and B. Faye. 2001. The use of multilevel models to evaluate sources of variation in reproductive performance in dairy cattle in Reunion Island. *Preventive Veterinary Medicine* 50: 127–144.

Elghafghuf, A., S. Dufour, K. Reyher, I. R. Dohoo, and H. Stryhn. 2014. Survival analysis of clinical mastitis data using a nested frailty Cox model fit as a mixed-effects Poisson model. *Preventive Veterinary Medicine* 117: 456–468.

Gröhn, Y. T., J. J. McDermott, Y. H. Schukken, J. A. Hertl, and S. W. Eicker. 1999. Analysis of correlated continuous repeated observations: modelling the effect of ketosis on milk yield in dairy cows. *Preventive Veterinary Medicine* 39: 137–153.

Martin, S. W., A. H. Meek, D. G. Davis, R. G. Thomson, J. A. Johnson, A. Lopez, L. Stephens, R. A. Curtis, J. F. Prescott, S. Rosendal, M. Savan, A. J. Zubaidy, and M. R. Bolton. 1980. Factors associated with mortality in feedlot cattle: the Bruce County Beef Cattle Project. *Canadian Journal of Comparative Medicine* 44: 1–10.

Martin, S. W., A. H. Meek, and P. Willeberg. 1987. *Veterinary Epidemiology: Principles and Methods.* Ames, IA: Iowa State University Press.

Martin, W. 1979. The Bruce County beef cattle project: factors related to mortality. *Canadian Veterinary Journal* 20: 336–337.

McDermott, J. J., D. M. Alves, N. G. Anderson, and S. W. Martin. 1991. "Benchmark"—a large observational study of Ontario beef breeding herds: Study design and collection of data. *Canadian Veterinary Journal* 32: 407–412.

McDermott, J. J., Y. H. Schukken, and M. M. Shoukri. 1994. Study design and analytic methods for data collected from clusters of animals. *Preventive Veterinary Medicine* 18: 175–191.

Reyher, K. K., S. Dufour, H. W. Barkema, L. Des Côteaux, T. J. Devries, I. R. Dohoo, G. P. Keefe, J. P. Roy, and D. T. Scholl. 2011. The National Cohort of Dairy Farms— a data collection platform for mastitis research in Canada. *Journal of Dairy Science* 94: 1616–1626.

Schukken, Y. H., Y. T. Grohn, B. McDermott, and J. J. McDermott. 2003. Analysis of correlated discrete observations: background, examples and solutions. *Preventive Veterinary Medicine* 59: 223–240.

Schwabe, C. W. 1977. *Epidemiology in Veterinary Practice*. Philadelphia: Lea & Febiger.

———. 1984. *Veterinary Medicine and Human Health*. 3rd ed. Baltimore: Lippincott Williams & Wilkins.

Smith, R. D. 2005. *Veterinary Clinical Epidemiology*. 3rd ed. Boca Raton, FL: CRC Press.

Stryhn, H., and J. Christensen. 2014. The analysis—Hierarchical models: Past, present and future. *Preventive Veterinary Medicine* 113: 304–312.

Thrusfield, M. 2007. *Veterinary Epidemiology*. 3rd ed. Oxford: Blackwell.

Wilesmith, J. W., G. A. Wells, M. P. Cranwell, and J. B. Ryan. 1988. Bovine spongiform encephalopathy: Epidemiological studies. *Veterinary Record* 123: 638–644.

7.5   About the author

Ian Dohoo is a professor of health management in the Atlantic Veterinary College at the University of Prince Edward Island, Charlottetown, Canada. He is a leading international figure in veterinary epidemiology and population-based health research. His extensive publication and graduate student supervision record, combined with authorship of the leading graduate-level text in the field (*Veterinary Epidemiologic Research*), have established his reputation. He led the development of the internationally recognized research program in veterinary epidemiology at the Atlantic Veterinary College and pioneered the development of computer-based animal health surveillance expertise in the region. He led the Colorado Veterinary Medical Association's expert panel on bovine growth hormone and has served as president of the Canadian Association of Veterinary Epidemiology and Preventive Medicine. He has received numerous teaching and research awards and is sought after internationally as a teacher of graduate-level epidemiology courses.

8 On learning introductory biostatistics and Stata

Marcello Pagano
Harvard T.H. Chan School of Public Health
Boston, MA

8.1 Introduction

The older we get, the more we reminisce. Mostly, these memories interest only the person imagining them, and possibly those featured in the memory and their families. Yet that has not stopped hundreds of biographers from practicing their craft. In that spirit, I add to the memory pile by revisiting a tiny slice of history from my perspective: namely, the enormous impact that computers in general, and Stata in particular, have had on biostatistics. Indeed, I would like to zero in on teaching in biostatistics and, moreover, on the teaching of biostatistics to health professionals who are not biostatisticians.

We start with me, taking my first statistics course as an undergraduate student more than 50 years ago. There were not yet any computational tools available to us then, so it took me an entire evening to draw my first histogram. Calculations became easier, quicker, and more accurate when we were introduced to a mechanical calculator—it was basically a Pascaline, like the odometer in your car but with a movable addend, just like the electric Monro-Matic that came later (http://webmuseum.mit.edu/detail.php?t=objects&type=browse&f=maker&s=Monroe+Calculator+Company&record=2) but with manual cranking (clockwise for addition and counterclockwise for subtraction). It was not until graduate school that I was exposed to the electric calculators that allowed us to complete analysis of variance calculations in "real time". Although students became quite adept at using these calculators—even inventing games on them to enable us to compete with one another—linear regression calculations became easy only once we were able to use the computer. This is when most any capable person became able to perform serious and substantive statistical calculations in a reasonable amount of time, especially once statistical packages were introduced. These packages were very expensive at first, especially the few that drew graphics, and they were full of bugs, often yielding untrustworthy results. This, of course, made us more vigilant of the results, which was not a bad thing. However, as calculating technology advanced, we were able to attack bigger and more complex problems and, therefore, able to practice statistics in more effective and interesting ways. The practice of statistics changed

61

because of the computer. We spent less time on calculations and more time on more complex calculations, on graphics, on mental "what if" experiments, and on much larger and varied datasets. This became even more so, later, when students started owning personal computers. Life was good.

This takes us to the late 1980s, which is when I started teaching a course, Principles of Biostatistics, assigned to students wishing to earn a degree (including the Master of Public Health degree) from the Harvard School of Public Health (now the Harvard T.H. Chan School of Public Health). Students can opt out of the course by taking an equivalent or more "advanced" course, yet knowledge of biostatistics is fundamental to the understanding of public health, so all graduating students must show some proficiency in biostatistics. Indeed, this knowledge has been a part of public health from day one.[1] But the challenge with the course is that it is not meant to teach students biostatistics as much as it is meant to help them become accomplished public health professionals. That the latter requires an appreciation for biostatistics, and for what it has to offer, is what must be conveyed to the student. Classically, this was achieved with formulas and calculators, but the advent of a package such as Stata allowed us to discard the inordinate amount of (sleep-inducing to nonstatisticians) time needed to demonstrate that

$$\sum_{i=1}^{n} \left(X_i - \overline{X}\right)^2 = \sum_{i=1}^{n} X_i^2 - n\overline{X}^2 \tag{8.1}$$

as well as to hope that package programmers had not used this formula. Do not misunderstand me. This is a beautiful formula that shows not only that the variance is positive, but also that it can be calculated in one pass through the data—something that is not obvious from looking at the left-hand side of the equation. With a little algebra, we can show that

$$\sum_{i=1}^{n} \left(X_i - \overline{X}\right)^2 = \frac{1}{2n} \sum_{i,j} (X_i - X_j)^2 \tag{8.2}$$

and relate the variance to the average squared distances between the observations, and further show that the variance can be restated without reference to the mean. (This last observation was pointed out to me by Nick Cox.) Of course, without algebra, we can show this empirically on many sets of numbers by using the computer, but that does not carry the certainty of a mathematical proof, nor the intuition obtained from deriving the formula and, parenthetically, some justification for dividing by $(n - 1)$.[2] But for this course, these relationships are usually (rightfully) not presented, and in that absence, possibly something is lost.

One property of Stata that I cannot overstate is its reliability. Harvard entered the free, massive online course offerings with two courses offered through edX (www.edx.org)

---

1. It is not a mere coincidence that the person in the United States who can lay claim to being the Father of Public Health in this country, Lemuel Shattuck, was also one of the founders of the American Statistical Association, and this introductory biostatistics course is as old as our school.
2. I have yet to introduce the formula for the variance without being asked why we divide by $(n - 1)$ and not n. My answer is inevitably, "Trust me. I am a doctor."

in 2012. One of the two was PH207X: Health in Numbers: Quantitative Methods in Clinical and Public Health Research, which is a combination of the Principles of Biostatistics and the introductory epidemiology course at the School of Public Health. Given that there were about 75,000 students who took part in the course, the school could not afford to pay the license fee for a statistical package. Of more importance, we did not have the personnel to handle the enquiries or complaints that would have come had we used a less reliable package. To help us, StataCorp allowed us to use Stata. The course went well.

8.2   Teaching the course

The three basic tenets of biostatistics are variability, inference, and probability (easily remembered by their acronym, VIP). These tenets form the basis for the course, and for good reason. We as statisticans are the only ones that can do these topics justice. These topics, of course, can be taught in several different ways, but one must remember that many of the students in the course are not interested in becoming biostatisticians. Indeed, by this stage of their education, most students have been conditioned by society to believe that they lack mathematical skills. Thus it seemed cruel and counterproductive to spend much time teaching the students various formulas for calculating means and variances and methods for drawing histograms, let alone teaching them how to calculate life tables, correlations, and regressions. This was the boring stuff—the plumbing—and there is so much poetry in inference, poetry to be revealed and appreciated. This is when I discovered Stata.

The first thing that impressed me about Stata was its graphics capability—I believe the cover of the Stata manual featured a wonderful scatterplot matrix. The histogram that previously took me all evening to draw, now took only a few minutes to set up, and just a few more seconds to draw or even to redraw with different settings. There were so many possibilities to use graphics to describe variability and relationships, though one must tread carefully before choosing a package for a course. Many packages have come and gone, so one should choose statistical software carefully before investing time and effort into learning how to use it. This is especially important when forcing a package onto students. Students trust their professors not to teach them skills that have a very short half-life. I chose Stata.

There were many reasons that I chose Stata, such as package capabilities, accuracy, affordability, and a promise of growth and advancement that would help maintain a high level of proficiency for the students. In short, Stata was of high quality. With time, that decision has also been reinforced by a track record that has convinced many others to use the package (note that the larger the user pool, the larger the number of testers able to debug the package and communicate with the developers, if need be). These users have not been shy in expressing their opinions—we have Statalist, a users' forum that is now more than twenty years old, with its loyal and active "Listers" who are surprisingly ready to generously and freely give their time to help other users. David Wormuth started Statalist with the help of Bill Mahoney, and after a year or so, I took over. Interestingly, StataCorp also uses Statalist to guide the developers. Stata also supplies an active help-line, 1-800-Stata-PC—a phone number I am still able to recall. I have fond memories of when, as a novice user, I would dial that number and Bill Gould—the president of the company—would answer my call! Unfortunately, and I say unfortunately because those conversations were delightful and extremely helpful, I do not believe he directly answers that phone number anymore.

After choosing to introduce the students to Stata—following expected resistance from administrative quarters who felt they knew better—the question became what difference, if any, did the program make to the course. I believe it completely transformed the character of the course. (I should add that I felt a little vindicated in my choice when Chelsea Clinton, Vice Chair of the Clinton Foundation, visited Harvard University for a leadership conference and exclaimed that she found statistics to be the most important part of her education [in terms of her current career] when earning her Master of Public Health from Columbia University. Then, when asked if she could program in Stata, she further exhibited her brilliance by responding, "Yea, I love Stata" [http://www.hsph.harvard.edu/voices/events/chelsea/]. This is the sort of evidence that works with administrators!)

To explain the impact that Stata has had on my teaching of biostatistics, I consider the three tenets one by one.

8.2.1   Variability

When teaching variability, we can easily calculate the mean and variance on practice sets, as we were accustomed to doing without computers, or on larger and more interesting real datasets if we have a computer to accurately do the drudge work. For an example of the latter, the National Heart, Lung, and Blood Institute, together with Boston University, provide a dataset from the Framingham Heart Study for public use (https://www.framinghamheartstudy.org/). This dataset, based on actual measurements from a very important study, contains many variables (around 70) on thousands of people (about 4,000) and tells an interesting story. (The data have been slightly jittered, to use a Stata phrase, to protect the privacy of the people involved. This altering does not impact the educational value of the dataset.) So with these data, the instructor and the students can calculate various means, standard deviations, and descriptive plots to start to appreciate variability as a whole, and even the range of variability across groups and time. Of course this contrasts with what many students have learned in the exact sciences in which variability is frowned upon because it often reveals sloppy measurement techniques. In this course, we teach that variability is not a bad thing. We focus on it to learn what it can help us understand about the phenomenon generating the data. But we have an uphill task. The term we have chosen, variability, suggests the synonym of irresponsibility, together with capriciousness, changeability, dizziness, fickleness, flippancy, frivolity, giddiness, and inconstancy. The list goes on. So why devote valuable time and effort to whimsy (yet another synonym) in a serious public health course? Being able to calculate a large number of standard deviations in various real-life situations, without having to ask students to dedicate unreasonable amounts of time to plodding arithmetic, allows us to appreciate the value of the variance and thus convince students of its value. For example, one can demonstrate the empirical rule by repeatedly showing the "truth" of the whimsical formula that, with some minor assumptions, the studentized variable is between minus and plus two roughly 95% of the time. This prepares students for the normal distribution. The added advantage is that now and in the future we will have even larger datasets to analyze, so students should become accustomed to dealing with large datasets and some of the problems to expect. In the past, we could not impose all of these calculations on students.

One command I found very powerful in Stata when I first discovered it—and it has always been there—is the lowly `by` command. This command is a useful tool for introducing students to variability as a function of another variable. It provides a good foundation for exhibiting patterns and for teaching regression later in the course. And, in conjunction with the command that calculates the variance, it helps provide an understanding of homo- and heteroskedasticity. When combined with boxplots, it also introduces the idea of prior distributions. This humble little command (which becomes even more powerful when combined with the `if` command), and the ability to use it almost universally, serves many uses and does not require very much time to explain. This consistency of commands—possibly the outcome of using one author for the original package—is a key feature of Stata that simplifies teaching students how to take advantage of the package. Unfortunately, because the package has matured and increased in size, there are now some inconsistencies, usually associated with multiple authors.

In the early days of this course, there was only enough time to teach students ways to calculate the mean and the variance. With the advent of Stata, the course can now develop a much greater appreciation for variability by the students, including the subtler aspects of the subject that we could not teach before (including the modeling of variability).

8.2.2 Inference

The advantages of teaching inference by using Stata are quite evident. The ability to easily sample repeatedly, either by simulation or by sampling from a big dataset such as the Framingham Heart Study (mentioned above), enables us to treat the dataset as a population and thus expose the arbitrariness in the definition of "population". Through this, we can teach the wonderful central limit theorem in action. The ability to clearly teach such concepts as a sampling distribution, or the difference between the standard deviation and the standard error, to nonmathematically inclined students (once an older student asked me what an integer was) is gratifying. I believe that our ability to reach such a large proportion of these students is largely due to our use of a statistical package—especially one like Stata that is not only powerful enough to use as an instruction tool, but is also easy enough for students to learn how to use briefly.

Whereas we could afford to teach only the formula for fitting a straight line before, we can now teach multiple linear regression, logistic regression, and even survival analysis, in the same amount of time. It was not possible before to reach so many minds and cover so much territory in one introductory course (around thirty hours of lectures and an equal amount of time in recitation). Granted, the students are no longer taught to do calculations with pencil and paper and instead are taught to call the computer via Stata only. As a result, they have lost the ability to act as a statistician if stranded on a desert island with no electricity. On the other hand, they have acquired a facility with the modern tools available to them, and they can spend more time on understanding the results and on the all-important diagnostic graphics. They have time to explore "what if" scenarios without the burden of hours of calculations. This is the poetry of the subject biostatistics. I cannot help but feel that learning the rhymes of the subject is a superior educational experience for these students as compared with spending their time on the routine of arithmetic operations. Many valuable tools are very well represented in Stata, and Stata's offerings expand with every new version (which is roughly every two years, according to the Nick Cox Frequency Law found on Statalist).

8.2.3 Probability

Teaching probability is much easier with a rich library of distributions to calculate or plot probability mass functions or to serve as a basis for simulations. Of course, previously we did not have to teach students how to calculate combinatorics to find binomial probabilities; we instead taught students how to look up values in a table. But that took time, too, and introduced a lot of errors into the calculations when students had to use probability tables, especially tables dealing with continuous distributions. This is now a much faster and easier process when using Stata. But the speed and accuracy with which these numbers can be calculated aside, I also believe that students are now exposed to a much broader and deeper understanding of probability. For example, we operate with a probability of less than 5% to define the rare event, yet the student must also appreciate the meaning of parts per million when dealing with a population of 300 million, as we do in public health. It is instructive to teach students that their chance of winning the lottery is approximately the same whether they buy a ticket or not, but explaining how big infinity is and demonstrating convergences (especially for rare events) is difficult without having a computer package capable of performing simulations. I have often wondered what the first book on probability would have looked like if Girolamo Cardano had used Stata.

8.2.4 Data quality

Now to discuss one last yet very important topic: data quality. Of course, Stata wields a large armamentarium to check for problems in the data, but typical introductory courses do not pay sufficient attention to this topic. Given the time constraints, one must choose what to include in a course—but we should not ignore the warts in any real set of data, and because the quality of our inference has an upper bound set by the quality of our data, some attention should be paid to the quality. Stata provides excellent data-management commands, so the program cannot be blamed for our lack of thoroughness. For instance, `max()` and `min()` provide range checks, and `if` provides logic checks for the easier checks. But students should also be taught (and Stata can help here, too) to search for patterns with missing data. For example, the all-important survey response rate is often overlooked to our own discredit, and Stata offers imputation software. Of course, this method is valid only when critical demanding assumptions are true, yet students can now be exposed to these computation-intensive methods, which was unthinkable in the past.

8.3   Conclusion

I think that the biostatistics we now teach is richer, more interesting, and more useful than the biostatistics we taught in the past, and this is largely thanks to Stata. The empirical justification for this is difficult to present. We could look at the student evaluations, and we have done well on that metric, but the popularity factor of the instructor is very much a part of that measurement. The measurement I value is the amount that students are able to accomplish after they have taken the course. How many return for more biostatistics courses, and how well do they do in those courses? How thoroughly do they realize the value of biostatistics in their research? We seem to have succeeded on all counts, so much so that even other fields are teaching more and more statistics as part of their own curricula!

The one concern I have is whether we are giving the key to a powerful package such as Stata to someone who has a false sense of confidence in its use. In other words, are we putting a twelve-year-old behind the wheel of a Ferrari? Of course this can happen with any tool, such as a scalpel, and we are able to insure ourselves through better training and an emphasis on the importance of assumptions for the valid use of any of these methods. Stata helps here, too, because it provides many tools to help train students and to assist in the examining of assumptions.

8.4   Acknowledgments

I thank Nick Cox and Marisa Pagano for improving this article. I thank all the teaching assistants who over the years have helped make teaching a joy, as well as the students who have made teaching a joy and a learning experience.

8.5   About the author

Professor Pagano obtained a PhD from Johns Hopkins University and has spent the last 35 years on the faculty at the Harvard T.H. Chan School of Public Health teaching biostatistics and advising students. His research in biostatistics continues to be on compute-intensive inference and surveillance methods that involve screening methodologies, with their associated laboratory tests, and in obtaining more accurate testing results that use existing technologies.

His interests extend to the quantitative aspects of monitoring and evaluation, especially as to how they are applied in resource-poor settings and used to improve the quality of health services to all.

9 Stata and public health in Ohio

Timothy Sahr
Ohio State University
Columbus, OH

9.1 Introduction

Since its inception, StataCorp has been committed to applied research methods in clinical and field aspects of public health through support of schools of public health and university-based summer programs. In Ohio, Stata has been used in the training of over 3,400 public health practitioners since 1998.

9.2 Stata distributed to Ohio counties

In 1998, to increase the public health workforce capacity of Ohio by increasing public health analytical skills throughout the state, the Ohio Department of Health and Ohio State University enabled the purchase of 150 copies of Stata software for local public health departments in 88 counties. To facilitate the use of quantitative software for public health practice, Ohio entities, in partnership with StataCorp, provided public health department epidemiologists and researchers with applied education emphasizing methods and techniques through a series of public health workforce development trainings named the Practice-Based Epidemiological Series. The pedagogical approach was to match public health theory to practice by incorporating field scenarios to core public health principles and practice techniques.

9.3 Local public health analytical personnel disparity

This partnership between public health agencies, academics, and Stata grew as a result of public workforce capacity assessments that indicated large skills variation between Ohio's local health departments. Research from the Association of Ohio Health Commissioners and Ohio Department of Health indicated that, since the 1980s, public health skills disparities existed between the larger and better financed counties and medium-to-smaller counties in terms of the number of field epidemiologists, public health researchers, and other analytical personnel. These assessments found that metropolitan and suburban counties were more likely to have adequate epidemiology, surveillance, and community risk assessment personnel than rural counties, which often lacked any personnel trained in epidemiology, biostatistics, health assessment, or determinants of health. Difficulty with the recruitment and retention of skilled public health professionals at local public health departments persist, with many health departments in medium- and small-sized counties lacking epidemiologists and public health researchers with graduate school training (for example, those holding a MPH, MS, or higher degree in public health or an associated discipline).

9.4 Training partnership with Stata

To address this uneven distribution of public health analytic skills, Ohio Department of Health's Data Committee, Ohio State University, StataCorp, and Ohio's public health partners decided to actively train the most promising of local public health department personnel in the core areas of public health scientific practice—emphasizing the areas represented within the National Public Health Performance Standards and areas reported as being of most concern to practicing epidemiologists and public health researchers. Core areas addressed included epidemiology, biostatistics, community-based public health research, community assessment, health disparities, risk evaluation, program development, and applied public health informatics. The faculties for these trainings were pulled from throughout the United States. Stata statistical software supported this workforce initiative in the areas of instruction, lab activities, and local practitioner support. To encourage analytical skills development in the counties, Stata was supplied to local county health departments. Stata was also provided to training participants for trial and discounted purchase.

9.5 Scope of trainings partnership

The Practice-Based Epidemiological Series seeks to build a stronger public health infrastructure by:

1. enhancing local health departments' key analytical personnel's scientific-based skill sets as they relate to community assessment, routine disease surveillance, disease outbreak surveillance and containment, public health emergency response, health systems data analysis, and survey research;

2. targeting all public health practitioners in Ohio with the goal of evening out the distribution of public health analytical skill sets between the large and medium or small counties;

3. fostering an applied cross-dialogue between field practitioners, Ohio health-associated state agencies, and the public health academic community; and

4. exposing attendees to standardized and cutting-edge methods and practices for addressing practice-based epidemiology issues.

9.6 Impact of the trainings

Past training evaluations and follow-up surveys indicate that: 1) Ohio's analytic and investigative public health workforce has been strengthened, with over 130 course units being taught to over 3,400 students; 2) the public health workforce's quantitative skills have been improved—particularly in the areas of epidemiology, biostatistics, health policy, and health system analyses; 3) Ohio's public health system has a stronger partnership with academic public health; and 4) public health accreditation requirements have been strengthened throughout the State.

9.7 Future goals

This applied workforce training will continue in the future with the help of StataCorp and its partners. As new public health issues present themselves, the training partnership is confident that quantitative analysis techniques for investigating public health will be offered and refined using available and future procedures found in Stata.

9.8 About the author

Timothy R. Sahr is the Director of Research and Analysis for the Ohio Colleges of Medicine Government Resource Center where he manages various research projects and policy initiatives concerning the health care system, behavioral risks, public health, environmental health, and the health status of Ohioans. Sahr's academic research interests include social epidemiological theory, adolescent behavioral risk, poverty studies, and religion and health. Tim has authored and coauthored numerous reports assessing health issues in Ohio communities. He has also authored and coauthored many professional briefs and academic articles in professional journals and has delivered lectures and presentations on various issues in public health, health policy, and religion and health. Sahr earned graduate degrees from Princeton and Ohio State, was an Honors Program student at Oxford University, and attended Anderson University. Currently, Sahr is also the coprincipal investigator for the 2015 Ohio Medicaid Assessment Survey, the Ohio Family Violence Prevention Project, the Primary Prevention Attitudinal and Awareness Survey, the Ohio Health Safety Net Research Project, the Health Records Integration Project, and various other projects associated with the Government Resource Center and Ohio Public Health Capacities initiatives. Sahr is the originator of the Practice-Based Epidemiology Series training for public health practitioners. Tim's prior employment included Director of Research at the Health Policy Institute of Ohio; Head of Policy at the Franklin County, Ohio, Board of Health; and survey researcher with Gallup International and Gallup Poll.

10 Statistics research and Stata in my life

Peter A. Lachenbruch
Oregon State University, retired
Corvallis, OR

10.1 Introduction

Throughout my career, I have used simulation heavily. For the first 20–25 years of it, I programmed simulations in FORTRAN. During that time, I had some misgivings about my own programming ability as well as FORTRAN's random-number generator. (Perhaps readers can note there should be other concerns!)

Many of the issues of my later work related to prediction problems—regression-like problems. These can include discriminant analysis as well as regression, and many of these problems involve selection of variables. The first tool that many statisticians use is stepwise regression. This procedure has many well-known shortcomings, yet it still is widely used. The selection of variables problem is quite similar. Early work compared the use of the stepwise procedure to selection of variables for univariate regressions. However, as the field progressed, researchers developed more sophisticated approaches. Nevertheless, many of the early approaches are still used today in spite of the newer work that has been published. Most of the poor approaches are used in fields in which the early work was learned by people who are now senior and have not kept up with statistics (they are typically not statisticians).

As for the more recent work, some of these newer methods are least angle regression (LARS) (see Lachenbruch [2011]) and least absolute shrinkage and selection operator (LASSO) (see Efron et al. [2004]). The abstract of the article by Efron et al. (2004) is as follows:

> The purpose of model selection algorithms such as all subsets, forward selection, and backward elimination is to choose a linear model on the basis of the same set of data to which the model will be applied. Typically we have available a large collection of possible covariates from which we hope to select a parsimonious set for the efficient prediction of a response variable. LARS, a new model selection algorithm, is a useful and less greedy version of traditional forward selection methods. Three main properties are derived:

1) A simple modification of the LARS algorithm implements the LASSO, an attractive version of ordinary least squares that constrains the sum of the absolute regression coefficients; the LARS modification calculates all possible LASSO estimates for a given problem, using an order of magnitude and less computer time than previous methods. 2) A different LARS modification efficiently implements forward stagewise linear regression, another promising new model selection method; this connection explains the similar numerical results previously observed for the LASSO and stagewise, and helps us understand the properties of both methods, which are seen as constrained versions of the simpler LARS algorithm. 3) A simple approximation for the degrees of freedom of a LARS estimate is available, from which we derive a Cp estimate of prediction error; this allows a principled choice among the range of possible LARS estimates. LARS and its variants are computationally efficient: the paper describes a publicly available algorithm that requires only the same order of magnitude of computational effort as ordinary least squares applied to the full set of covariates.

The LASSO summary (Tibshirani 1996) is as follows:

We propose a new method for estimation in linear models. The LASSO minimizes the residual sum of squares subject to the sum of the absolute value of the coefficients being less than a constant. Because of the nature of this constraint, it tends to produce some coefficients that are exactly 0 and hence vies interpretable models. Our simulation studies suggest that the LASSO has some favorable properties of both subset selection and ridge regression. It produces interpretable models like subset selection and exhibits the stability of ridge regression.

These methods provide a theoretical basis for selection of variables. The Stata routines that implement these do not allow for a tuning parameter to stop selection at an early time, so these methods tend to select more variables than a stepwise regression that has a small p to enter or a large p to remove. However, a user can avoid this problem by selecting only those variables that enter at a specified level of significance (that is, ignore the least significant selected variables). LARS and LASSO are performed using the same Stata command (based on an implementation in R). Discriminant analysis can be thought of as a regression on binary variables (for example, a logistic regression). In one sense, it is like estimation with missing values. The old approaches of complete cases or mean imputations (both discarded) are still used, and, similarly, the last observation carried forward is still kaput.

10.2   My career using Stata

In the early years of my career, I worked on the effects of misclassification in the training samples for discriminant analysis and on estimating the error rates (Lachenbruch

1966, 1967; Lachenbruch and Mickey 1968). Here I developed the leave-one-out method, which successively omits one observation, recalculates the discriminant function, and classifies the omitted observation . A bit of algebra makes this possible to do without multiple matrix inverses. My work on this subject was done using FORTRAN programs. These programs were written by me, and it is likely that they had some bugs. However, Stata's simulation commands use Stata's own random-number generator and rely only on programming the discriminant rules, which is less error prone than the earlier FORTRAN code.

In 1988, I became interested in improving the performance of diagnostic tests. I considered how to combine a second sample from a set of patients, thus resampling the data used to create the diagnostic tests (that is, taking a second or third sample from each patient). This would lead to a smoothing of the error rate estimates. I then programmed Stata for this. I wrote an article (Lachenbruch 1988) that examined the effects of various multiple reading procedures on sensitivity and specificity. Multiple reading procedures entailed the performance of a diagnostic test several times and the consequent assignment of the subject to an "affected" or "unaffected" group. A unanimity rule (that is, all tests must be positive) led to the largest predictive value positive, but it may have had unacceptably low sensitivity. The rule that classified the subject as positive based on most of the tests increased both the sensitivity and specificity of the individual test. Variability of subjects' sensitivity and specificity (that is, correct diagnosis on a test) affected the performance of these rules. I also studied alternative procedures in which the final test was a better but more expensive test.

In the early 1990s, I became concerned with the effects of unequal variance for the Wilcoxon test and the Kruskal–Wallis test (Lachenbruch 1991; Lachenbruch and Clements 1991). I had an application that clearly had unequal variances, and I wanted to be sure that the tests had the proper size and did not lose much power (this was for the Wilcoxon test but I broadened the scope to the Kruskal–Wallis). I found that the tests held their size and did not lose a lot of power. If I were to do this again today, I would probably use a permutation test on the ranks or just on the means.

In the mid-1970s, I had a consulting problem that concerned the growth of organisms on agar—some plates showed growth while others did not. I modeled this problem as a two-part model that consisted of a binomial part (growth/no growth) and a linear regression part to model the growth, conditional on growth being present. About 15 years later, a doctoral student had a problem modeling hospitalization insurance expenditures—at the time, about 95% of enrollees in plans had no hospitalization in a year. This was another example of a two-part model (however, the doctoral student did not take my advice). Other examples abound, I found that many researchers used a normal model or a Wilcoxon test for this sort of data. I then decided to look at several competitors of the two-part models (Lachenbruch 2001a). These included the Wilcoxon, the normal, etc. I found that, in some circumstances, the Wilcoxon was as good as the two-part model, but in other cases, the two-part model was better. My work led me to make recommendations on power and sample-size requirements (Lachenbruch 2001b). I completed this work using the simulation routines in Stata. An interesting sidelight is that a two-part model implies identifiability of the 0 values. If this is not the case, then a mixture model should be used.

In a problem involving a product approval for a rare disease, the sponsor noted that the response should have a Poisson distribution. The Food and Drug Administration was not sure of this suggestion, so we checked to see if a Poisson distribution was needed by using a variance test (Lachenbruch et al. 2001). This was important to check, because the primary analysis was changed from a Wilcoxon to a Poisson regression, and the p-value went from 0.27 to 0.0004. The sponsor then noted that they did not need to invoke the Poisson model but could use a permutation test. In our analyses, we found that the assumptions were not met (there were several outliers). When we used a permutation test, the p-value went to 0.57. The high p-values were associated with cases in which the outliers were assigned to the treatment group. We then assessed the reference that the sponsor used to justify the Poisson and found that the data were far from a Poisson (the test of the variance was highly significant).

Another time in my career, a fellow statistician asked me if I knew how transforming to normality affected inferences. I did not, so I examined the Box–Cox transformation and how it changed the size and power of two sample tests. I compared these results with permutation tests (Lachenbruch 2004). Note that there are several methods besides the t test and Wilcoxon test. One of the most direct methods is to perform a bootstrap or a permutation test on the data. This is a simple nonparametric test. In my article on this, I examined the p-values from the t test and the Wilcoxon test on both the original and transformed data. Of course, doing the Wilcoxon on the transformed data gave the same results, because ranks are invariant under monotonic transformations. I found that only distributions with high asymmetry or heavy tails seriously affect the t test. The Box–Cox likelihood ratio test appeared to have some advantages over the other tests, but this is offset by the greater complexity in making the results understandable to nonstatisticians. Overall, I concluded that the variability in outcomes with the different procedures demonstrates the importance of specifying such procedures a priori.

During my career, I found that I often wanted to get two-way tables of one variable against a set of others. Gutierrez and I (mostly Bobby) developed a modification of the `tabulate` command (Gutierrez and Lachenbruch 2009). In many research contexts, we had a number of categorical variables and wanted to look at contingency tables for many variables. We often had one variable that categorized a variable of particular interest. For example, we wanted to study if domestic or foreign cars differ on these variables. Thus we were interested in the three tables `mpgcat`, `cost`, and `rep78` versus `foreign`. In this case, the `firstonly` option to `tab2` allowed us to get the contingency table of the first versus the others.

```
. tab2 foreign mpgcat cost rep78, firstonly chi2
```

In one study, I desired to select variables that would predict survival. I did this with a `foreach` loop, but I thought it would be nice to have an option to the `tabulate` command. In my study, there were 145 observations and 25 variables. Many variables had missing values, and I was concerned that these could affect which variables were selected. When I tried a regression, the missing values reduced the dataset used for the multiple regression to 76—that is, there were only 76 complete cases. To improve prediction, I generated 10 variables that were complete and independent of the 25 (denoted as u-variables). These variables should have had no impact on the prediction. Using stepwise regression with a p-to-remove of 0.2, Stata selected 16 variables, including 4 of the u-variables. When I reduced the p-to-remove to 0.1, only 9 variables remained selected. I also used LARS regression, and this selected 14 variables with 2 u-variables. I considered several imputation methods—I could replace the missing values with the mean of the nonmissing values, use the last observation carried forward, or use multiple imputation (with stepwise regression). Overall, some of the variables were continuous (although skewed), some were dichotomous, and some were categorical and ordered. The percent missing ranged from 0 to 19, with most being complete. In the end, I found that multiple imputation coupled with stepwise selection of variables worked well. This held for LARS as well as stepwise regression. The LARS method performed similarly to the LASSO method.

Stata's built-in routines speed up the research process and greatly assure correctness. These examples from my research demonstrate how Stata can help. Much present work still relies on simulation. One persistent problem with simulation has been inadequate sample size. A sample with only 100 or 500 cases is generally inadequate for simulation problems but can usually be handled with a simple sample-size calculation, so there is no excuse for too-small studies.

References

Efron, B., T. Hastie, I. Johnstone, and R. Tibshirani. 2004. Least angle regression. *Annals of Statistics* 32: 407–499.

Gutierrez, R. G., and P. A. Lachenbruch. 2009. Stata tip 74: firstonly, a new option for tab2. *Stata Journal* 9: 169–170.

Lachenbruch, P., L. Marzella, W. Schwieterman, K. Weiss, and J. Siegel. 2001. Poisson distribution to assess actinic keratoses in xeroderma pigmentosum. *Lancet* 358: 925.

Lachenbruch, P. A. 1966. Discriminant analysis when the initial samples are misclassified. *Technometrics* 8: 657–662.

———. 1967. An almost unbiased method of obtaining confidence intervals for the probability of misclassification in discriminant analysis. *Biometrics* 23: 639–645.

———. 1988. Multiple reading procedures: The performance of diagnostic tests. *Statistics in Medicine* 7: 549–557.

———. 1991. The performance of tests when observations have different variances. *Journal of Statistical Computation and Simulation* 40: 83–92.

———. 2001a. Comparisons of two-part models with competitors. *Statistics in Medicine* 20: 1215–1234.

———. 2001b. Power and sample size requirements for two-part models. *Statistics in Medicine* 20: 1235–1238.

———. 2004. Proper metrics for clinical trials: Transformations and other procedures to remove non-normality effects. *Statistics in Medicine* 22: 3823–3842.

———. 2011. Variable selection when missing values are present: A case study. *Statistical Methods in Medical Research* 20: 429–444.

Lachenbruch, P. A., and P. J. Clements. 1991. Anova, Kruskal–Wallis, normal scores and unequal variance. *Communications in Statistics—Theory and Methods* 20: 107–126.

Lachenbruch, P. A., and M. R. Mickey. 1968. Estimation of error rates in discriminant analysis. *Technometrics* 10: 1–11.

Tibshirani, R. 1996. Regression shrinkage and selection via the LASSO. *Journal of the Royal Statistical Society, Series B* 58: 267–288.

10.3 About the author

Peter Lachenbruch grew up in Los Angeles and attended University of California, Los Angeles as an undergraduate. He then attended Lehigh University for a Master's degree in mathematics, after which he attended the University of California, Los Angeles for a PhD in biostatistics (1965). He then went to University of North Carolina, where he advanced to professor in 1975. He soon moved to the University of Iowa, where he remained until 1986 before moving to University of California, Los Angeles to be a professor of biostatistics. He met Bill Gould during this period and began using Stata. His work included studies in rheumatology and psychiatric epidemiology, among others. He has taught using Stata for about 30 years. His work has resulted in over 200 publications in various journals. Lachenbruch has been chair of the American Public Health Association's statistics section, the president of the Biometric Society (Eastern North American Region), and president of the American Statistical Association. He has been retired since 1975.

11 Public policy and Stata

Stephen P. Jenkins
Department of Social Policy
London School of Economics and Political Science
London, UK

11.1 Introduction

I was invited to write about how public policy has evolved over the past three decades and how Stata has been part of this process. This is an impossible brief, so I am going to be selective in terms of coverage and, even then, all perspectives provided are strongly colored by my own career as an applied economist, which, as it happens, spans roughly the same three decades as Stata's. I limit my scope to the areas of health, education, welfare, and the labor market, as well as to individuals, families, households, and statistical analysis of survey or administrative data. This is a "micro" perspective; I am not discussing macroeconomics or time-series data. My subject is quantitative policy analysis rather than the public policies themselves. I am focusing on topics researched primarily by social scientists, and mainly those in which economists and econometricians now play an influential role.

11.2 The credibility revolution in public policy analysis

Let me begin by putting on the hat of a contemporary mainstream empirical microeconomist. From this perspective, there is a very clear view of what has happened to policy analysis over the last three decades: there has been substantial change in approach, and it has all been for the better. As Angrist and Pischke put it, "[E]mpirical microeconomics has experienced a credibility revolution, with a consequent increase in policy relevance and scientific impact. [...] the primary engine driving improvement has been a focus on the quality of empirical research designs" (2010, 4).

By empirical research designs, Angrist and Pischke mean methods and datasets that allow analysts to identify causal effects credibly, referring to approaches based on random assignment to treatment and control groups (randomized control trials) and to natural and quasi-experiments. In all the approaches, the researcher seeks data in which there is variation across cases in a key treatment variable, and that variation can be taken as exogenously given. Regression-based methods in the quasi-experimental approach include instrumental variables, regression discontinuities, or differences in differences. (In the latter case, fixed-effects estimators applied to panel data are commonly used to control for time-invariant unobserved confounders.) Nonparametric methods for comparing treated and control cases that account for observable differences are based on covariate adjustment using matching by propensity score, nearest neighbor, kernel, or related reweighting methods.

What has been Stata's role in these developments? I believe it has been substantial for two reasons. The first is that, even though Stata is not essential to implement the purer experimental evaluations, many researchers have used it anyway because it was already their software of choice for data management and it also had the generic statistical tools required. The closer the research design is to a randomized control trial, the closer the estimation of treatment effects is to a simple comparison of means. Therefore, the statistical component of the evaluation is a relatively straightforward task compared with getting the empirical research design and data right.

The second reason Stata has played a substantial role in these developments is that it made specialist statistical routines for estimating treatment effects widely available to users early in the credibility revolution. Randomized control trials are relatively rare in evaluations of public policy because of perceived ethical or infeasibility problems (it is difficult to randomly assign marital status or, say, differences in drug or alcohol consumption), or because of high set-up costs and lack of specialist know-how. The events facilitating natural experiments are also relatively rare. Furthermore, external validity is an issue for both types of experiment. Therefore, quasi-experimental designs based on observational data have been the most prevalent approach and, in this case, statistical analysis takes on a greater role in the overall evaluation. Over the last decade or so, Stata introduced many commands that provided the requisite specialist tools and made them widely available to users.

The leading example is the `psmatch2` package by Edwin Leuven and Barbara Sianesi (2003). This package implemented not only Mahalanobis and propensity-score matching, but also integrated tools for checking for common support and covariate imbalance. The `psmatch2` command, which was built on Sianesi's (2001) `psmatch` package, was first released in the Statistical Software Components (SSC) archive at Boston College in April 2003 and has been frequently updated since then. This package is currently ranked number one for total software downloads from the SSC (18,257 downloads as of 6 July 2014; http://logec.repec.org/scripts/paperstat.pf?h=RePEc:boc:bocode:s432001). (Less well-known is the suite of programs providing similar functionality that accompanies Becker and Ichino's [2002] *Stata Journal* article. Abadie et al. [2004] provided extensions focusing on nearest-neighbor matching.) It was only in 2013 that the number of `psmatch2` downloads began to fall, undoubtedly because of the release of Stata 13 with its `teffects` suite of estimators, though the number remains relatively large.

Another Stata package important for policy evaluation is `ivreg2` (Baum, Schaffer, and Stillman 2003, 2007) and its panel-data sibling, `xtivreg2` (Schaffer 2011). These do instrumental variables regression for linear models, providing functionality beyond that of Stata's built-in command `ivreg` and its successor `ivregress`. `ivreg2` is ranked fourth for total downloads from SSC (first released May 2002; 14,657 downloads as of

6 July 2014), and `xtivreg2` is ranked eighth in total downloads (first released November 2005; 7,746 downloads as of 6 July 2014). Software for regression discontinuity analysis was provided by Austin Nichols's `rd` package (14th on the SSC total download list with 14,657 downloads since November 2007 as of 6 July 2014). Although basic differences-in-differences analysis can be straightforwardly implemented using Stata's built-in commands, some extensions require more specialized estimators, some of which are provided by Juan Miguel Villa's `diff` package (17th on the SSC total download, ranking with 4,956 downloads since October 2009 as of 6 July 2014).

Although the SSC is not the only source of Stata code for policy analysis and the usefulness of download statistics can be questioned, my summary judgment is that the SSC data provide good prima facie evidence of Stata's contribution to policy analysis in the postcredibility revolution environment.

11.3 A broader view of what counts as valuable public policy analysis

There is more than one way to assess what counts as public policy analysis. Although the credibility revolution of contemporary mainstream empirical microeconomics is very influential (and rightly so), it incorporates a rather narrow view of valuable policy analysis. There is also a substantial contribution made by what some researchers label rather disparagingly as "descriptive" analysis.

I would contend that knowing how things are, or how they have changed or compare with another country, is an essential prerequisite to any discussion about policy options and priorities, let alone any sort of evaluation of specific policy measures. Some of the biggest policy debates are founded on arguments about "the facts". Recent examples in the UK include how much intergenerational social mobility there is and how it has changed over time; how much real incomes, inequality, and poverty rates have changed in the aftermath of the Great Recession in the era of austerity; the educational performance of school children and how it differs by ethnic minority group, social background, and type of school attended; employment rates and earnings of recent migrants compared with native-born UK workers; differences in the inflation rates faced by pensioners and other groups, etc.

As soon as the definition of policy analysis is broadened to include this sort of research, Stata's role will be considered to be even more significant than it already is. This is because the people doing this sort of work are increasingly using Stata to do their analysis; the user base widens beyond credible revolutionaries and extends to many other quantitative researchers in universities and increasing numbers of people working in research institutes and local, national, and international governmental agencies.

How has descriptive policy analysis changed over the last three decades? One fundamental change concerns the data available and the capacity to analyze it. Policy-relevant quantitative research requires good data. Let me take the UK as an example. Thirty years ago, unit-record data from household surveys were only just beginning to become more widely accessible to researchers, facilitated by the work of the UK Data Archive acting as a national data library. However, analysis of such data was constrained by hardware. The survey data files were considered "large" and typically held on magnetic tape and accessed via a mainframe computer. At this time, the personal computer had only recently been invented and few could be found in universities. Statistical software packages were limited in their functionality and their integration. (I began my career using one package for data management and another for estimation.) The teaching of econometrics often focused on time-series analysis, reflecting the availability of such data. There were no national longitudinal surveys. Fast forward to 2014, and the world is totally different.

Public policy analysts now have access to a vast amount of data from multiple sources. Household survey data are routinely available from national data archives or downloadable from the Internet, and there is a wealth of longitudinal as well as cross-sectional survey data. Administrative data are increasingly part of researchers' portfolios too, and there are many more possibilities for data combination through linkages across data sources, often using geo-referenced identifiers. We can easily access data from other countries in addition to our own. The Internet itself is providing data, and the era of Big Data is upon us.

Our networked personal computers have the capacity to not only store data files that are much larger than could have been imagined thirty years ago but also to analyze them using a vastly extended portfolio of statistical tools. For example, methods for analysis of limited dependent variables, panel and survival data, sample selection, robust standard errors, and survey-design effects are routinely available. The training of graduate social scientists in quantitative methods has improved in parallel.

As a result of these developments, commissioners and funders of research expect more in terms of coverage of data (for example, drawing on a full time series of cross-sectional surveys, rather than simply one or two) and statistical sophistication, and researchers deliver more. However, one thing that has not changed is the report style required for nonacademic audiences such as national and international agencies and some research foundations. Reports must still be written in accessible, plain English for readers without the quantitative skills of the researchers. Communicating research findings effectively to nonacademic audiences is as big a challenge to policy analysts today as it was thirty years ago. It may even be a greater challenge now because the gap between the statistical training and experience of researchers and nonacademic research users has probably widened.

Stata has played a significant part in the descriptive policy analysis context, although its role is hard to quantify precisely—especially in a manner that would satisfy credibility revolutionaries! One strong example of Stata's role is its take-up among researchers who do quantitative policy analysis. I am not privy to StataCorp's sales figures, but it is manifest that, whereas the market for general statistical software was dominated by two Goliaths thirty years ago, Stata is the David of today. Moreover, casual observation suggests that Stata is increasingly being used to train the policy analysts of tomorrow in the quantitative methods teaching in universities today.

Stata use has increased substantially not only in universities and nongovernmental research institutes but also in governmental agencies with research capacity. Internet searches on *"agency name Stata"* frequently lead to evidence of Stata use, especially when the agency is North American or an international organization such as World Health Organization or the World Bank. National statistical offices have tended to favor common software approaches, standardizing on a single relational database management system interfaced with Some Alternative Software. But even in these environments, Stata is increasingly used for specialist tasks. (For Canadian and UK examples, see McCrosky [2012] and Barnes [2002].)

Early adoption of Stata by researchers with esteem and influence, together with free sharing of resources of multiple complementary kinds, has led to an increase in take-up and use by others—a form of virtuous circle. A leading example of this is the role played by the World Bank's research department and associated researchers from outside the bank. They were early adopters of Stata and have long made programs freely available to outside researchers. One example is the collection of stand-alone programs in the "Poverty Analysis Toolkit" (http://go.worldbank.org/YF9PVNXJY0). This has recently been substantially redeveloped (extending the topic coverage) and is now part of an integrated software environment, ADePT (Automated DEC Poverty Tables), which is built on top of Numerics by Stata—a version of Stata that is embedded within applications developed by others. See Poi (2010) for further discussion.

ADePT is freely downloadable from http://go.worldbank.org/UDTL02A390, together with extensive documentation and datasets. The latest version contains eight modules for the analysis of poverty, inequality, social transfers, labor, gender, health equity, education, and food security. Accompanying these are five books, all freely downloadable, that introduce the underlying analytical methods with extensive examples written by leaders in the respective fields. The quality and comprehensiveness of *A Unified Approach to Measuring Poverty and Inequality* (Foster et al. 2013), combined with its unbeatable price, is such that I plan to adopt it as a course textbook in the coming academic year.

Earlier World Bank books have had substantial influence in their fields. See, for example, *Analyzing Health Equity Data Using Household Survey Data* (O'Donnell et al. 2008), which includes a large collection of Stata examples and downloadable resources. The pioneer par excellence is Deaton's magisterial *The Analysis of Household Surveys* (1997, 1998), a source I still consult and also use in teaching. Deaton, one of the world's leading economists (and 2009 President of the American Economic Association), wrote in his preface that in his experience, "[Stata] is the most convenient package for working with data from household surveys" (1997, 2). Furthermore, a feature unusual for its time, the book provided the Stata code for the analysis, thereby enabling others to implement the methods, many of which were advanced then and not easily available elsewhere.

11.4   What is it about Stata?

So, Stata has made a substantial and growing contribution to quantitative public policy analysis over the last thirty years. In many ways, the features of Stata that underpin this contribution are the same characteristics that make it the software of choice for other forms of quantitative analysis. What are these?

It is tempting to begin by simply pointing to the way Stata integrates tools for data management, statistical analysis, and graphics; however, that hardly makes it distinctive among competing software packages and, arguably, my preferences for Stata over others could reflect habit. Nevertheless, I would point to two aspects of Stata that have been particularly important in my own research career. First, when I switched to Stata as my main statistical software in around 1994, I was struck by its emphasis on do-files and log files and, hence, the ability to create audit trails and reproducible results. As a researcher and research project team manager, I rate this capability very highly. As a creator of data released to a wider public (income variables for the British Household Panel Survey), it was essential. Second, Stata shifted to the new suite of graphics commands in version 8 (2003). With this update, not only could chart creation be automated using do-file code, but there were substantial improvements in functionality per se. Earlier, I referred to the need for policy analysts to effectively communicate their work to research users, and my experience is that well-designed graphs are particularly valuable for this.

There are six factors underpinning Stata's success in addition to its integrated nature. Here follows a reprise of my list from a decade ago (Jenkins 2005), suitably updated. First, there is Stata's extensibility—building in the capacity for users to extend Stata themselves (using ado-files and, more recently, Mata), combined with an openness and encouragement to do so from StataCorp. Second, there was early exploitation of the Internet, with seamless integration of the ability to download free software updates and user-written programs and to search for such materials. Third, Stata runs in RAM memory and is relatively fast. Although memory capacity was once a constraint, it was recognized early on that the ever-falling price of memory meant that this would not remain a practical problem. Fourth, Stata is produced for virtually all operating systems and is the same for each one.

Fifth, StataCorp fosters close links with its users—it listens. For example, it sends staff to the independently run Stata user group meetings worldwide, and developers present scientific papers, run short courses, and host "wishes and grumbles" sessions with users. Stata developers read and contribute to Statalist. StataCorp provides users with technical support of a quality that is unparalleled.

Sixth, and perhaps of most vital importance to researchers, Stata does not sacrifice academic integrity—Stata is for science. It provides capacity for cutting-edge statistical methods, but it does so in a suitably conservative manner. Implementation of methods typically follows scientific acceptance (as with the treatment effects packages cited earlier), often based on consultations with specialist experts in the relevant field (again, as with the treatment effects packages cited earlier), and always comes after extensive in-house validation and certification exercises. Stata's ties with science have been fostered by the development of the *Stata Journal* and the publication of many excellent "[...] with Stata" books by world-leading econometricians and statisticians.

What links all these features is a type of integration and connectedness between software, developers, and users that constitutes a virtuous circle that has led to mutual advantage over the last three decades and shows no sign of abating. Stata will continue to play a major role in quantitative analysis of all kinds and in policy analysis in particular. Happy 30th birthday, Stata!

11.5 Acknowledgments

I owe personal thanks to Bill Gould and all at StataCorp for their contribution to my research career: Stata has been an ever-present companion in my work over the last quarter century. Specific thanks to Vince Wiggins for suggestions about how to proceed with this essay and to Lucinda Platt for comments on the first draft.

References

Abadie, A., D. Drukker, J. Leber Herr, and G. W. Imbens. 2004. Implementing matching estimators for average treatment effects in Stata. *Stata Journal* 4: 290–311.

Angrist, J. D., and J.-S. Pischke. 2010. The credibility revolution in empirical economics: How better research design is taking the con out of econometrics. *Journal of Economic Perspectives* 24: 3–30.

Barnes, M. 2002. Using Stata at the Office for National Statistics. UK Stata Users Group meeting proceedings. http://www.stata.com/meeting/8uk/ONSstatausergroup.pdf.

Baum, C. F., M. E. Schaffer, and S. Stillman. 2003. Instrumental variables and GMM: Estimation and testing. *Stata Journal* 3: 1–31.

———. 2007. Enhanced routines for instrumental variables/generalized method of moments estimation and testing. *Stata Journal* 7: 465–506.

Becker, S. O., and A. Ichino. 2002. Estimation of average treatment effects based on propensity scores. *Stata Journal* 2: 358–377.

Deaton, A. 1997. *The Analysis of Household Surveys: A Microeconometric Approach to Development Policy.* Baltimore: Johns Hopkins University Press for the World Bank. http://documents.worldbank.org/curated/en/1997/07/694690/analysis-household-surveys-microeconometric-approach-development-policy.

————. 1998. deaton: Stata modules to analyze household surveys. Statistical Software Components S360701, Department of Economics, Boston College. http://ideas.repec.org/c/boc/bocode/s360701.html.

Foster, J. E., S. Seth, M. Lokshin, and Z. Sajaia. 2013. *A Unified Approach to Measuring Poverty and Inequality.* Washington, DC: The World Bank.

Jenkins, S. P. 2005. Oration for the award of honorary doctorate to William Gould. Colchester, UK: University of Essex. http://www.essex.ac.uk/honorary_graduates/or/#2005.

Leuven, E., and B. Sianesi. 2003. psmatch2: Stata module to perform full Mahalanobis and propensity score matching, common support graphing, and covariate imbalance testing. Statistical Software Components S432001, Department of Economics, Boston College. http://ideas.repec.org/c/boc/bocode/s432001.html.

McCrosky, J. 2012. Custom Stata commands for semi-automatic confidentiality screening of Statistics Canada data. Presentation at 2012 Stata Conference, San Diego. http://www.stata.com/meeting/sandiego12/materials/sd12_mccrosky.pdf.

Nichols, A. 2007. rd: Stata module for regression discontinuity estimation. Statistical Software Components S456888, Department of Economics, Boston College. http://ideas.repec.org/c/boc/bocode/s456888.html.

O'Donnell, O., E. van Doorslaer, A. Wagstaff, and M. Lindelow. 2008. *Analyzing Health Equity Data Using Household Survey Data.* Washington, DC: The World Bank.

Poi, B. 2010. Stata makes a difference at the World Bank: Automated poverty analysis. http://www.stata.com/news/statanews.25.2.pdf.

Schaffer, M. E. 2005. xtivreg2: Stata module to perform extended IV/2SLS, GMM and AC/HAC, LIML and *k*-class regression for panel data models. Statistical Software Components S456501, Department of Economics, Boston College. http://ideas.repec.org/c/boc/bocode/s456501.html.

Sianesi, B. 2001. Implementing propensity score matching estimators with Stata. UK Stata Users Group meeting proceedings. http://fmwww.bc.edu/repec/usug2001/psmatch.pdf.

Villa, J. M. 2009. diff: Stata module to perform differences in differences estimation. Statistical Software Components S457083, Department of Economics, Boston College. http://ideas.repec.org/c/boc/bocode/s457083.html.

11.6   About the author

Stephen P. Jenkins is Professor of Economic and Social Policy at the London School of Economics and Political Science and was formerly at the Institute for Social and Economic Research, University of Essex, where he was director 2006–2009. Stephen has been a Stata user since version 2.1, he is the author of multiple frequently downloaded commands on SSC, and he is a regular presenter of Stata-based short courses on survival analysis and statistical graphics. He is an associate editor of the *Stata Journal*, and he has helped organize the UK Stata User Group meetings in London every second year since 1999. He has had a long career in policy research, including projects for UK government departments and agencies, New Zealand Treasury, and the OECD. Stephen is currently the editor-in-chief of the *Journal of Economic Inequality*.

12 Microeconometrics and Stata over the past 30 years

A. Colin Cameron
Department of Economics
University of California–Davis

12.1 Introduction

Microeconomics research has become much more empirically oriented over the past 30 years. This has been made possible by increased computational power. The IBM XT 286, introduced in 1986, had 640KB of RAM, a 6MHz processor, a 20MB hard disk, and a 1.2MB processor.[1] By contrast, a typical PC today runs more than 500 times faster, with memory and storage that is more than 10,000 times larger. This greater computer power has been accompanied by increased data availability, new methods, and the development of statistical software to implement these methods.

Here I discuss how theoretical and applied microeconometrics research has evolved over the past 30 years and how Stata has been part of this process. The discussion of theory is necessarily brief, with further detail provided in Cameron (2009). The role of Stata, one of several packages available to econometricians, is especially important because it is now the most commonly used package in applied microeconometrics.

The interplay between theory and implementation is not straightforward because considerable time can pass from the introduction of new methods to their use by applied researchers and their incorporation in a statistical package. This delay is partly due to it taking time before the usefulness of the new method is clear. To some extent, this is a "chicken and egg problem" because methods are used much more once they are incorporated into a statistical package. Delay also arises because some methods—notably semiparametric regression, maximum simulated likelihood, and Bayesian methods—are difficult to code into a user-friendly command that will work for a wide range of problems. Because Stata is programmable, the speed of this process can be (and has been) accelerated by users developing their own code ahead of any official Stata command. In some cases, this code is made available to other Stata users as a user-written Stata ado-file. Here I mention only a few of these useful add-ons.

---

1. See http://en.wikipedia.org/wiki/IBM_Personal_Computer_XT.

12.2   Regression and Stata

Many of the core regression methods now widely used in applied microeconometrics research were introduced in the late 1970s and early 1980s. These methods include sample selection models (Heckman 1976), quantile regression (Koenker and Bassett 1978), bootstrap (Efron 1979), heteroskedastic-robust standard errors (White 1980, 1982), and generalized method of moments (GMM) estimation (Hansen 1982). Additionally, several seminal books appeared in the early and mid-1980s, namely, *Limited-Dependent and Qualitative Variables in Econometrics* (Maddala 1983) for limited dependent variable models, *Advanced Econometrics* (Amemiya 1985) for nonlinear regression models, and *Analysis of Panel Data* (Hsiao 2003) for panel data.

Cox (2005) provides a brief history of the first 20 years of Stata; Baum, Schaffer, and Stillman (2011) provide a recent overview. Stata was introduced in 1985 for use on IBM PCs running under DOS rather than on a mainframe computer. The initial release of Stata was quite limited and focused primarily on tools for data management and exploratory data analysis due to both its newness and the low computing power of PCs. The only regression command in the initial release was command `regress` for least-squares estimation of linear models.

The basic limited dependent variable models were among the first regression models to be introduced into Stata—logit and probit in 1987, survival models in 1988, tobit models and multinomial logit models in 1992, and linear sample selection models and negative binomial models in 1993. Quantile regression methods became much more widely used after their incorporation in Stata in 1992. Commands for general nonlinear least-squares and maximum likelihood estimation were introduced in 1993. GMM estimation was incorporated in several linear model commands, though a general command for GMM estimation was not introduced until 2009. The basic panel-data commands, a strength of Stata, were introduced in 1995 (linear) and 1996 (nonlinear).

Increased computing power has enabled greater use of simulation methods. Monte Carlo experiments using a known data-generating process can be conducted in Stata via the command `simulate` or the command `postfile`. Random variables can be drawn directly from a multitude of distributions following a major Stata enhancement in 2008. These distributions include the multivariate normal and the truncated multivariate normal (using the Geweke–Hajivassiliou–Keane simulator). The Stata random-number generators include Halton and Hammersley sequences in addition to a standard random-uniform generator.

Methods for simulation-based estimation of parametric models were developed in the 1980s and 1990s, especially maximum simulated likelihood estimation (McFadden 1989; Pakes and Pollard 1989) and Bayesian Markov chain Monte Carlo methods (Geman and Geman 1984). These methods have enabled the estimation of increasingly complex parametric models. In empirical microeconometrics, these are most often limited dependent variable models such as the random parameters logit model. Furthermore, Bayesian methods are generally used merely as a tool; the results are still given a frequentist interpretation rather than a Bayesian interpretation.

Stata initially introduced Bayesian methods in particular contexts, notably with the command `asmprobit`, which estimates the multinomial probit model using maximum simulated likelihood, and with multiple imputation commands that use Markov chain Monte Carlo methods. And user-written code provided Stata front ends to the Bayesian statistical packages WinBUGS (Thompson, Palmer, and Moreno 2006) and MLwiN (Leckie and Charlton 2013). The command `bayesmh`, introduced in Stata 14, may lead to much greater use of Bayesian methods.

Stata avoids use of simulation-based estimation methods whenever possible. In particular, complex parametric models are often difficult to estimate because of an intractable integral. For a one-dimensional integral, such as that in the linear random-effects model, it is standard to use Gaussian quadrature rather than simulation methods. For higher dimensional integrals of the multivariate normal that appear in mixed models, Stata commands `mixed` and `gsem` use adaptive multivariate Gaussian quadrature rather than simulation methods.

An alternative strand of research has developed methods to estimate regression models that rely on relatively weak distributional assumptions. The building block is nonparametric regression on a single regressor. Several methods have been proposed in statistics literature, beginning with kernel regression in 1964, followed by lowess, local polynomial regression, wavelets, and splines. Stata initially provided lowess estimation. Local polynomial regression, including kernel regression and local linear as special cases, appeared as command `lpoly` in 2007. These nonparametric regression commands and the kernel density estimation command `kdensity` are especially valuable for viewing data and key statistical output such as residuals.

The single-regressor nonparametric regression methods do not extend well to models with multiple regressors because of the curse of dimensionality. Econometricians have been at the forefront of developing semiparametric models that combine a high-dimensional parametric component with a low-dimensional (usually single-dimensional) component. The late 1980s and early 1990s saw development of estimation methods for three commonly used models—the partial linear model, the single-index model, and generalized additive models. Semiparametric methods are particularly useful for limited dependent variable models with censoring and truncation because they enable crucial parametric assumptions on unobservables to be weakened; Pagan and Ullah (1999) provide a survey. These semiparametric methods generally require selection of smoothing parameters, sometimes with deliberate undersmoothing or oversmoothing. Perhaps this is why there are no official Stata commands for semiparametric regression, though there are some Stata add-ons for some specific estimators. The lack of semiparametric regression commands in Stata is one reason that semiparametrics methods (a focus of recent theoretical econometrics research) are infrequently used in applied microeconometrics.

In addition to obtaining regression coefficients under minimal assumptions, the econometrics literature has developed methods for statistical inference under minimal assumptions. Heteroskedastic-robust standard errors were developed by White (1980, 1982) and introduced into Stata 3 in 1992. If model errors are clustered, then default and heteroskedastic-robust standard errors can be much too small. Extensions to cluster–

robust inference were made by Liang and Zeger (1986) and Arellano (1987). Including cluster–robust standard errors (Rogers 1993) in basic Stata regression commands early on greatly increased Stata's usage. Even though Stata is at the forefront in providing robust standard errors, the inclusion of a cluster–robust option for the more advanced estimation commands took considerable time.

When standard errors (nonrobust or robust) are not available, they can be obtained by using an appropriate bootstrap. A `bootstrap` command appeared in Stata in 1991 with significant enhancement in 2003. The theoretical literature has emphasized a second use of the bootstrap, namely, bootstraps with asymptotic refinement that may lead to better finite-sample inference. These latter bootstraps are seldom used in practice; a notable exception is the wild-cluster bootstrap when there are few clusters (Cameron, Gelbach, and Miller 2008). Bootstraps with refinement, such as bias-corrected confidence intervals as a bootstrap option and other methods with some additional coding, can also be implemented in Stata.

A distinguishing feature of econometrics is the desire to make causal inference from observational data. Instrumental-variables estimation and its extension to GMM were the dominant methods when Stata was introduced. Articles by Nelson and Startz (1990) and Bound, Jaeger, and Baker (1995) highlighted the need for alternative inference methods when instruments are weak. Recent results on weak instrument asymptotics for linear models with nonindependent identically distributed model errors, the usual case in empirical microeconomics studies, are implemented in Stata add-ons `ivreg2` (Baum, Schaffer, and Stillman 2007) and `weakiv` (Finlay, Magnusson, and Schaffer 2013).

A major change in causal microeconometrics research is the use of the potential outcomes framework of Rubin (1974) that has evolved into the quasi-experimental or treatment-effects literature, summarized in Angrist and Pischke (2009). Matching methods such as propensity-score matching (Rosenbaum and Rubin 1983) or use inverse-probability weighting can be used for selection on observables only. A Stata command to implement these methods was introduced in 2013 and superseded earlier user-written add-ons. When selection is also on unobservables, most methods can be implemented using existing Stata commands. These methods include local average treatment-effects estimation (Imbens and Angrist 1994), a reinterpretation of instrumental variables when treatment effects are heterogeneous, fixed-effects panel models and their extension to differences-in-differences with repeated cross-section data, sample-selection models, and regression discontinuity design. For dynamic linear panel models with fixed effects, the methods of Arellano and Bond (1991) and extensions can be implemented using the official Stata command `xtabond` and the user-written add-on `xtabond2` (Roodman 2009).

Methods for spatially correlated data have been progressively developed over the past 30 years. Currently there are no official Stata commands for spatial regression, but there are several user-written Stata add-ons that handle and analyze spatial data, including the spatial regression command `sppack` (Drukker et al. 2011).

Researchers in biostatistics and in social sciences other than economics, who are also Stata users, use some regression methods that are not often used in empirical microeconometrics. Generalized linear models (command `glm`) and generalized estimating equations (`xtgee`) cover many nonlinear regression models, including those with binary or count dependent variables. Mixed models or hierarchical models (command `mixed`) can lead to more precise estimation than a simple random-effects model can when model errors are clustered. Other social sciences make greater use of completely specified structural models (commands `sem` and `gsem`).

12.3   Empirical research and Stata

There is more to empirical research than obtaining parameter estimates and their standard errors.

The first step of empirical research is to simply analyze and view the data ahead of any regression analysis. Useful graphical methods are kernel density estimates and two-way scatterplots with a fitted nonparametric regression curve. Stata introduced a very rich publication-quality graphics package in 2003. Interpreting the sources of variation in grouped data is simplified by using the `statsby` command and `xt` commands such as `xtsum`, `xttab`, and `xtdescribe`.

Model diagnostics and specification tests can be useful. Applied microeconometrics studies tend not to use available methods that can detect outlying observations and influential observations. This is in part due to concerns about subsequently overfitting a model, though such diagnostics can also highlight mistakes such as miscoded data. Available model specification tests are infrequently used, notable exceptions being Hausman tests and tests of overidentifying restrictions. Stata postestimation commands include these standard methods, and they also enable in-sample and out-of-sample prediction.

Many applied studies in microeconometrics seek to estimate a marginal effect, such as the increase in earnings with one more year of schooling, rather than a regression model parameter per se. Marginal effects and their associated standard errors can be computed using the `margins` command introduced in 2009 that supplanted the user-written command `margeff` (Bartus 2005). Factor variables, also introduced in 2009, enable extension to models with interacted regressors.

Empirical microeconomics studies are increasingly based on data sources that are very complex. Complications include: 1) data may come from several different sources; 2) data may come from surveys; 3) data may have a grouped structure such as panel data or individual-level data from several villages; and 4) some data may be missing.

A real contribution of Stata has been its ability to handle these complications. Stata is a data-management package, in addition to a statistical package, with features including the ability to handle string variables and commands to merge and append datasets. The Stata survey commands control for weighting, clustering, and stratification. Empirical microeconometrics studies generally do not use the survey commands. Instead, regular estimation commands are used with weights, if necessary, and with appropriate cluster–robust standard errors. Stratification is ignored, with some potential loss in estimator efficiency. Grouped data can be manipulated using the `by` prefix command and the `reshape` command. Stata's estimation commands automatically allow for missing data using case-deletion. If case-deletion is not valid, then weighted regression can be used if weights are available. Alternatively, one can use the Stata multiple-imputation command introduced in 2009. For imputation, empirical economics researchers currently rely on case-deletion or on crude imputation methods such as hot-deck imputation, despite their limitations.

Stata was initially limited in the size of dataset it could handle because it requires that all data be stored in memory to speed up computations. This limitation has greatly diminished over time given increases in computer memory capacity and the emergence of 64-bit PCs.

As empirical studies have become more complex, the need for replicability has increased. Researchers need to be able to keep track of their own work, return to it after leaving it for a considerable period of time, and potentially coordinate computations with coauthors, research assistants, and students. Furthermore, several leading journals require that data and programs be posted at their archives. Stata is well suited for producing replicable studies because it is command driven, and the resulting Stata scripts can be run on a wide range of platforms and on newer versions of Stata.

As is clear from the previous section, it can take considerable time before a new method is included in a statistical package such as Stata. It is, therefore, advantageous to use software that is programmable. Stata has always been programmable, and it includes the complete matrix programming language Mata that was introduced in 2007.

The widespread use of Stata has created a community of users. Stata encourages this community through the Stata Technical Bulletin (which began in 1990 and was superseded by the *Stata Journal* in 2001), Statalist Server (1994), Stata Users Group meetings (1995), the Stata website (1996), and Stata Press books (1999). For basic applied microeconometrics, the books by Baum (2006), Cameron and Trivedi (2010) and Mitchell (2012) are especially helpful. The websites for introductory econometrics texts provide code for analysis in Stata. The Statistical Software Components (1997) website provides many Stata user-written programs that can be directly downloaded to Stata. As already noted, Stata users have provided many useful add-on programs. While some have been superseded by official Stata commands, many still fill gaps or augment official Stata commands.

As with any statistical package, the ubiquity of Stata also has downsides. Data analyses may be restricted only to what is easily implemented in Stata. Researchers may not understand the limitations of the methods used, such as tobit model estimates relying on very strong parametric assumptions. Also Stata may eventually become legacy software, yet one with a very large user base. To date, Stata has avoided this by continuing to target academic researchers in economics, other social sciences, and biostatistics.

References

Amemiya, T. 1985. *Advanced Econometrics*. Cambridge: Harvard University Press.

Angrist, J. D., and J.-S. Pischke. 2009. *Mostly Harmless Econometrics: An Empiricist's Companion*. Princeton, NJ: Princeton University Press.

Arellano, M. 1987. Computing robust standard errors for within-groups estimators. *Oxford Bulletin of Economics and Statistics* 49: 431–434.

Arellano, M., and S. Bond. 1991. Some tests of specification for panel data: Monte Carlo evidence and an application to employment equations. *Review of Economic Studies* 58: 277–297.

Bartus, T. 2005. Estimation of marginal effects using margeff. *Stata Journal* 5: 309–329.

Baum, C. F. 2006. *An Introduction to Modern Econometrics Using Stata*. College Station, TX: Stata Press.

Baum, C. F., M. E. Schaffer, and S. Stillman. 2007. Enhanced routines for instrumental variables/generalized method of moments estimation and testing. *Stata Journal* 7: 465–506.

———. 2011. Using Stata for applied research: Reviewing its capabilities. *Journal of Economic Surveys* 25: 380–394.

Bound, J., D. A. Jaeger, and R. M. Baker. 1995. Problems with instrumental variable estimation when the correlation between the instruments and the endogenous explanatory variables is weak. *Journal of the American Statistical Association* 90: 443–450.

Cameron, A. C. 2009. Some recent developments in microeconometrics. In *Palgrave Handbook of Econometrics. Volume 2: Applied Econometrics*, ed. T. C. Mills and K. Patterson, 729–774. London: Palgrave Macmillan.

Cameron, A. C., J. B. Gelbach, and D. L. Miller. 2008. Bootstrap-based improvements for inference with clustered errors. *Review of Economics and Statistics* 90: 414–427.

Cameron, A. C., and P. K. Trivedi. 2010. *Microeconometrics Using Stata*. Rev. ed. College Station, TX: Stata Press.

Cox, N. J. 2005. A brief history of Stata on its 20th anniversary. *Stata Journal* 5: 2–18.

Drukker, D. M., H. Peng, I. Prucha, and R. Raciborski. 2011. sppack: Stata module for cross-section spatial-autoregressive models. Statistical Software Components S457245, Department of Economics, Boston College. http://ideas.repec.org/c/boc/bocode/s457245.html.

Efron, B. 1979. Bootstrap methods: Another look at the jackknife. *Annals of Statistics* 7: 1–26.

Finlay, K., L. M. Magnusson, and M. E. Schaffer. 2013. weakiv: Stata module to perform weak-instrument-robust tests and confidence intervals for instrumental-variable (IV) estimation of linear, probit and tobit models. Statistical Software Components S457684, Department of Economics, Boston College. http://econpapers.repec.org/software/bocbocode/s457684.htm.

Geman, S., and D. Geman. 1984. Stochastic relaxation, Gibbs distributions, and the Bayesian restoration of images. *IEEE Transactions on Pattern Analysis and Machine Intelligence* 6: 721–741.

Hansen, L. P. 1982. Large sample properties of generalized method of moments estimators. *Econometrica* 50: 1029–1054.

Heckman, J. J. 1976. The common structure of statistical models of truncation, sample selection and limited dependent variables and a simple estimator for such models. *Annals of Economic and Social Measurement* 5: 475–492.

Hsiao, C. 2003. *Analysis of Panel Data*. 2nd ed. Cambridge: Cambridge University Press.

Imbens, G. W., and J. D. Angrist. 1994. Identification and estimation of local average treatment effects. *Econometrica* 62: 467–475.

Koenker, R., and G. Bassett, Jr. 1978. Regression quantiles. *Econometrica* 46: 33–50.

Leckie, G., and C. Charlton. 2013. runmlwin—a program to run the MLwiN multilevel modelling software from within Stata. *Journal of Statistical Software* 52: 1–40.

Liang, K.-Y., and S. L. Zeger. 1986. Longitudinal data analysis using generalized linear models. *Biometrika* 73: 13–22.

Maddala, G. S. 1983. *Limited-Dependent and Qualitative Variables in Econometrics*. Cambridge: Cambridge University Press.

McFadden, D. 1989. A method of simulated moments for estimation of discrete response models without numerical integration. *Econometrica* 57: 995–1026.

Mitchell, M. N. 2012. *A Visual Guide to Stata Graphics*. 3rd ed. College Station, TX: Stata Press.

Nelson, C. R., and R. Startz. 1990. The distribution of the instrumental variables estimator and its t-ratio when the instrument is a poor one. *Journal of Business* 63: S125–S140.

Pagan, A., and A. Ullah. 1999. *Nonparametric Econometrics*. Cambridge: Cambridge University Press.

Pakes, A., and D. Pollard. 1989. Simulation and the asymptotics of optimization estimators. *Econometrica* 57: 1027–1057.

Rogers, W. H. 1993. sg17: Regression standard errors in clustered samples. *Stata Technical Bulletin* 13: 19–23. Reprinted in *Stata Technical Bulletin Reprints*, vol. 3, pp. 88–94. College Station, TX: Stata Press.

Roodman, D. M. 2009. How to do xtabond2: An introduction to difference and system GMM in Stata. *Stata Journal* 9: 86–136.

Rosenbaum, P. R., and D. B. Rubin. 1983. The central role of the propensity score in observational studies for causal effects. *Biometrika* 70: 41–55.

Rubin, D. B. 1974. Estimating causal effects of treatments in randomized and nonrandomized studies. *Journal of Educational Psychology* 66: 688–701.

Thompson, J., T. Palmer, and S. Moreno. 2006. Bayesian analysis in Stata with WinBUGS. *Stata Journal* 6: 530–549.

White, H. 1980. A heteroskedasticity-consistent covariance matrix estimator and a direct test for heteroskedasticity. *Econometrica* 48: 817–838.

———. 1982. Maximum likelihood estimation of misspecified models. *Econometrica* 50: 1–25.

12.4 About the author

Colin Cameron is a Professor of Economics at University of California–Davis. He is the coauthor with Pravin Trivedi of three gradual-level monographs on microeconometrics and on count data. His current research focuses on cluster–robust inference.

13 Stata enabling state-of-the-art research in accounting, business, and finance

Demetris Christodoulou
University of Sydney
Sydney, Australia

13.1 Introduction

This essay makes the case that Stata is the software of choice for top research in accounting, business, and finance. Indeed, to some extent it is argued that Stata has helped accelerate the application of the state-of-the-art methods found in the high-impact journals.

Hand-collected data on software citations from 530 scholarly peer-reviewed journals that are classified as part of accounting, business, and finance research areas suggests that Stata is cited more times than any other software in high-impact journals. Impact is defined as the top quartile of the SCImago Journal Rank (SJR) factors among these journals. Stata citations in top journals refer to the application of state-of-the-art methods, specifically on advanced panel-data estimation, standard-error correction, survival analysis, Heckman selection models, propensity-score matching, multinomial modeling and more recently complex time series analysis, generalized method of moments (GMM), and other relevant methods. It is evident from the data that Stata has lived up to its original motto that it is indeed a *"Serious software for serious researchers"*.[1]

Considering the lagged effect of upgrades, the data suggest that Stata's takeover of market share started with the release of version 7, where it established the software as a quality provider of data management, panel-data methods, and survival analysis. StataCorp's strategy to overhaul interface and modernize the look of the software with the massive release of version 8 (bigger, faster, graphical user interface (GUI) interface, new graphics engine) and the bonus release of Mata with version 9 widened its customer base from the point-and-click novice to the most advanced object-oriented programmer. The concurrent release of additional advanced panel-data estimators, sophisticated time-series analysis tools, a complete maximum-likelihood estimation suite, GMM, and other

---

1. This is Stata's old trademarked motto, currently replaced by the more conforming "Data analysis and statistical software".

modern tools cemented Stata's presence in financial research. Another key event that was timed very strategically in the midst of the great recession was the free release of the voluminous PDF documentation.

Given the data, Stata's overall growth rate is only second to SPSS, which is purposefully designed to cater to the masses (SPSS appears in quantity at the lower end of journal rank impact).[2] SAS's growth in scholarly research across the three areas seems to be mediocre and to have stagnated during 2007–2009. MATLAB, RATS, and Eviews have a strong presence only in finance, but with even lower growth than SAS. There is no data available on the usage of the R-project software because it was impossible to search for the keyword "R".

Stata has many challenges lying ahead, and the obvious one is how to cope with the ever-growing size of data. Stata is designed to place data in memory, and the question is whether computing innovation can keep up with the growth in data as accounting, business, and finance research are increasingly drawing links with very large data.[3] Also it is surprising that Stata has yet to sort out the issue with publication-quality tables and rotating 3D graphs (the perennial requests at the Stata Users Group meetings).

In this essay, I also draw lessons from my personal journey with Stata as a researcher and teacher in financial analysis and quantitative methods and as a consultant in data analysis for industry and government.[4] I feel honored to be invited to contribute to the celebratory volume on the 30 years of Stata and only wish to be here again for the next round of celebrations.

13.2   Data on software citations and impact

To determine the extent of software usage over time in accounting, finance, and business, I observed the number of software citations in scholarly journal articles. After examining many articles, it appears that statistical software is generally cited in journals for the following reasons: i) compliance to editorial policy or specific request by reviewers, or ii) description or promotion of new methods applied in that area or that particular journal. The two factors are often related. Thus, the analysis on software citations is largely an analysis on software usage in state-of-the-art methods.

---

2. SPSS is also the language of choice by a large margin for publications appearing in open-access journals and conference proceedings, particularly in business.
3. CSC claims that data production in 2020 will be 44 times greater than in 2009, with 70% of this data produced by individuals, particularly from Internet traffic and mobile devices (http://www.csc.com/insights/flxwd/78931-big_data_universe_beginning_to_explode).
4. Since 2008, I have provided training to more than 1,000 executives, researchers, and PhD students on data analysis using Stata.

The protocol for data hand-collection is outlined below, with criteria listed in order of priority:[5]

- *Scholarly*: observe only journals that are classified as scholarly peer-reviewed journals with clear blind-reviewing processes. Do not observe trade journals or magazines, which are identified as such by their publishers or the digital libraries. Do not observe open-access journals as identified by the publishers or the Directory of Open Access Journals.[6]

- *Inclusiveness*: include all journals that are classified by publishers as part of the broadly defined accounting, finance, or business areas. In addition, search for journals that contain in their titles the following keywords: accounting, audit, bank, bankrupt, bond, business, corporate governance, credit, derivative, enterprise, finance, financial, forecasting, insurance, insolvency, investment, pension, portfolio, real estate, risk, tax, and trusts.

- *Searchability*: include journals with online searchable facilities. Thoroughly search the databases of notable journal publishers: Cambridge, De Gruyter, Elsevier, Emerald Insight, InderScience, Oxford, Palgrave, SAGE, Taylor & Francis, Wiley, and relevant professional associations. Scan the digital libraries of EBSCO, JSTOR, PROQUEST, and ScienceDirect for relevant journals and content.

- *Representation*: search for citations of notable statistical software: GenStat, GLIM, MATLAB, Minitab, NLOGIT, Octave, RATS, SAS, SHAZAM, SPSS, Stata, Systat, and TSP. The most notable omission is the R-project due to the commonality in its keyword (that is, the single letter "R"). No search was performed for Microsoft Excel because it is not marketed as a statistical software.

- *Near exactness*: search for keywords as exact phrases, that is, within double quotes. For instance, the search for "RATS" gives results for this exact word, and not other words that may contain the sequence r-a-t-s. Filter false results, for example, "SAS" also stands for Statement of Auditing Standard, Scandinavian Airline Systems, Security Association of Singapore, self-assessment system, selective attention strategy, and more. When in doubt and where there is access, download the PDF document and perform a thorough search. Given the volume of hand-collected data, results cannot be exact, but false positives or misses are unbiased and the margin of error should be immaterial.

- *Measure of impact*: high impact is defined as the top quartile of the SJR among the accounting, finance, and business journals—the cutoff point is SJR equal to 0.94.[7] As a sensitivity check, results remain robust when high impact is defined at the top quartile of the Thomson Reuters impact factor (TRIF). There are 279

---

5. The data have been hand collected during July–August 2014.
6. See http://doaj.org. Open-access journals are highly controversial because many have been accused of substandard reviewing processes, or even with predatory scope where authors pay to publish. I could not find data to help me establish the quality of open-access journals, hence, their total exclusion.
7. The SJR measures the scientific influence of the average article in a journal and expresses how central to the global scientific discussion is an average article of that journal (http://scimagojr.com).

journals (out of 530) with nonmissing SJR factors, and the TRIF confirms that most of these journals have very low impact factors.

The *Appendix* presents a table with the number of software citations for 530 scholarly journals that are classified as part of the accounting area [**A**], the business area [**B**], or the finance area [**F**]. Data are reported only for software with more than 200 citations across all journals during 1999–2013. In order of total number of citations, these software are SPSS, Stata, SAS, MATLAB, RATS, Eviews, S-Plus, and Limdep. The table also gives the SJR factors.

Figures 13.1, 13.2, and 13.3 give the timeline charts of total software usage independent of impact for the accounting, business, and finance areas, respectively, as reported in the *Appendix*. The shaded area notes the period of the great recession (or "global financial crisis"). The labels on the top x axis indicate the years of Stata upgrades (versions). Figure 13.1 shows that Stata has overtaken SAS in the accounting area with the release of version 12 and that it has narrowed the margin with SPSS significantly. Figure 13.2 suggests that Stata still has a lot of convincing to do in the business research area, where it is placed as the second most-preferred statistical software overall but with a very wide margin from SPSS. Figure 13.3 shows how Stata is the preferred software overall in finance, and that the lead was established with the release of version 11, which together with version 12 was largely focused on time-series analysis upgrades.

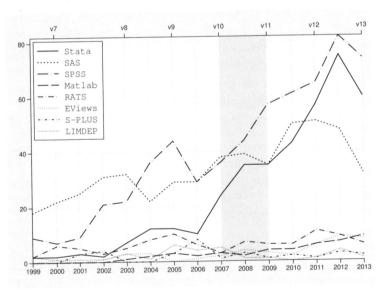

Figure 13.1. Total software citations in scholarly research journals: Accounting area

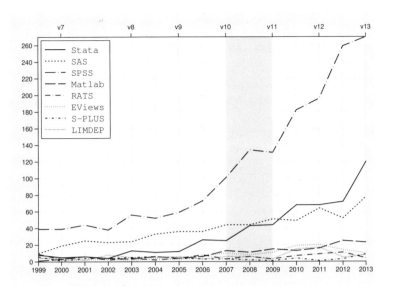

Figure 13.2. Total software citations in scholarly research journals: Business area

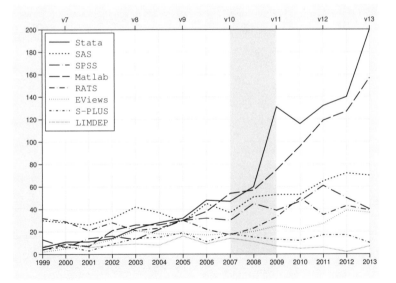

Figure 13.3. Total software citations in scholarly research journals: Finance area

Figure 13.4 presents quantile–quantile plots of the log-transformed number of software citations times the SJR factor across the three areas for journals with nonmissing SJR. The measure aims to capture the rate of software usage weighted by journal impact. The quantiles of $\ln$ (Stata citations $\times$ SJR) are plotted against the quantiles of $\ln$ (SAS citations $\times$ SJR) for low-impact journals classified as the first three quartiles of SJR (left-hand-side plot) and separately for high-impact journals classified as the top quartile of SJR (right-hand-side plot).[8] The same graphs are produced for Stata versus SPSS and Stata versus MATLAB. The message is clear—Stata is the software of choice for research published in high-impact journals in all areas.

---

8. SJR quartiles are highlighted by SCImago as a classification of impact. The visual analysis is robust when "high impact" is defined at the top quartile of the TRIF.

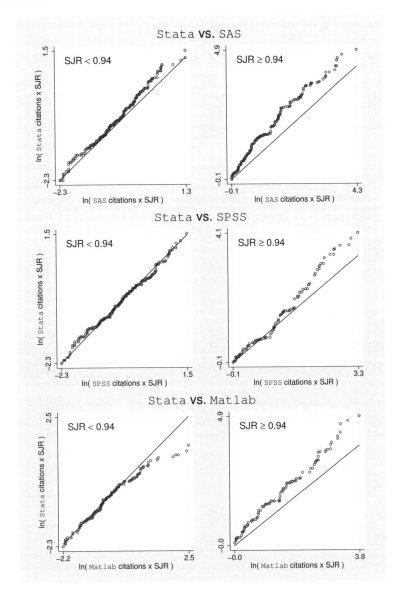

Figure 13.4. Quantile–quantile plots: software citations by journal impact

13.3   Earning market share

Statistical software usage tends to be sticky in that once you invest in learning a language, you tend to accumulate experience and not move on to some other platform. It takes courage from the user and a lot of convincing from the software provider to switch languages and, as the data show, Stata's strategy for gaining market share has worked exceedingly well specifically in the areas of accounting and finance (figures 13.1 and 13.3), but less so in business (figure 13.2).

For most, the choice of statistical software begins in the early training years and is defined by the following factors, presented in order of significance: i) availability of licenses and institutional training, ii) availability of packaged methods in GUI or command-line interfaces (CLIs), iii) influence of advisors, iv) software documentation and books for novices, v) flexibility and intuitiveness of language, and vi) peer pressure.

There is also the argument of industry pressure, that is, that the choice of software is also determined by what is considered to be the standard in that industry. Specifically, finance research students are still being told that working in the financial industry requires knowledge of SAS for data analysis or MATLAB for financial engineering applications and simulation. Also accounting and business research students are being convinced that data-driven industry requires at least knowledge of Excel and SPSS and that universities should cater for these needs.[9] This may have been true fifteen years ago (before Stata 8), but today this perception is no longer relevant.

13.3.1 Stata 7

Stata has always presented an attractive platform for the serious researcher, but I would identify version 7 as the single most defining point for when Stata entered the race for market share in accounting, business, and finance.

My story with Stata also begins with version 7, and it would strike as familiar to many looking back at their PhD training years. At the time, I traversed across several statistical platforms, and it was the norm to explore software depending on packaged GUI or CLI capabilities.[10] Software training was scarce, and there was lack of advanced textbooks on software applications. The rather painful experience offered no control over the quality of the analysis.

---

9. I can say that even at the University of Sydney Business School (the oldest and one of the largest Australian universities) these expectations are strongly enforced in our graduate programs. Our accounting and business graduates learn Excel and perhaps also SPSS, and our finance graduates learn SAS. Stata is taught only in PhD courses because of my involvement in the MEAFA research group. There is no reason why Stata cannot replace SPSS, SAS, or MATLAB as the recommended Business School statistical software given the scope of work in graduate studies.

10. During the first two years of my PhD training, I explored Eviews, GLIM, Limdep, LISREL, MATLAB, Minitab, OxMetrics, and SAS, and this was considered to be normal at the time.

I was attracted to Stata 7 because of its panel-data capabilities and the quality of the accompanied documentation—the very appealing [XT] manual. My first-time experience was a revelation. The syntax was intuitive, like speaking to some other person. The documentation was complete and assumed zero knowledge, but quickly ascended to advanced applications with thorough breakdown of estimation procedures. Panel-data analysis was no longer a black box, and I could even see how to reproduce the canned results through manual calculations. This was the first time I had full control of the analysis. I was also surprised to find out that, contrary to my peers' misleading perceptions, Stata was a champion in data management.

13.3.2 Interface and code overhaul

Stata 8 was a massive upgrade. The aim was to redesign the code and modernize the interface to make the software relevant to a wider customer base. Stata 8 was bigger, faster, and more attractive. Its GUI (drop-down menus and dialog boxes) appealed to the point-and-click crowd and allowed StataCorp to tap into the pool of SPSS and Excel-type users, but with a difference: Stata's GUI is in tune with its CLI and allows the novice to learn syntax by exploring the familiar environment of the drop-down menus.[11]

Stata 8 also offered a revamp of the graphics engine, bringing its visualization capabilities up to par with other competing platforms. However, I must say that sometimes it feels that the graph code has been written by a different person than the rest of the software. But Nick Cox's contribution in making graphics accessible must be acknowledged and, in this respect, he maintains a legendary status in Statalist. Also his *Stata Journal* tips have been instrumental in making sense of the vast graph language. Stata 10 completed the graphics overhaul with the introduction of the interactive Graph Editor that again was aimed to the point-and-click user. I understand the strategic purpose of the Graph Editor, but this feature seems to be at odds with Stata's fundamental philosophy that it is important to learn how to code because any interactive edits are not interpreted into command lines.

The release of Mata with Stata 9 appealed to a completely different type of user. Mata is a full-fledged matrix programming language, and anyone with experience in MATLAB, Ox, or Gauss would find this extremely appealing. Mata fully communicates with Stata and can be used interactively, or Mata functions can be developed to be called from within Stata. Mata is much faster and is best for computationally intensive work.[12]

---

11. Before this release, new users were advised to go through the help files and type-a-bit to get-a-bit. However, with Stata 8, the approach switched to using the drop-down menus to click-a-bit to get-a-bit. Regardless the approach, the ultimate aim is to transfer sequence commands in do-files and execute from there.

12. I remember Bill Gould raising some eyebrows at the Australian Users Meeting in Adelaide in 2006 when he claimed that Mata is as fast as MATLAB, and probably faster. We had to try to believe it.

13.3.3   Longitudinal panel-data analysis

Stata's foresight to invest heavily in panel-data analysis remains as one of the most determining factors for earning market share in accounting, business, and finance. Research in these areas is defined by repeated observations over time, and panel-data methods are the obvious choice for analysis. Pooled-data analysis (that is, disregard of time and cross-sectional variations) was once accepted as standard, but as panel-data methods became widely available, today it is panel-data analysis that is considered to be the norm.

Stata 7 and 8 provided the first complete suite of panel-data commands including static, dynamic, and nonlinear estimation. In this respect, David Roodman's work on `xtabond2` has been instrumental in attracting a wave of new users. The *Stata Journal* article by Roodman (2009) on `xtabond2` has been cited more than 2,300 times.[13]

The highlight of Stata 9 was the `xtmixed` command, which could fit many-way linear mixed or multilevel models, including random coefficients models. `xtmixed` opened new possibilities in financial research. Stata 9 also refined existing panel-data commands, including the `robust` and `cluster` variance corrections that are now standard in most applications.

Stata 10 extended panel-data capabilities to binary and count responses (`xtmelogit` and `xtmepoisson`), and it introduced the all-encompassing dynamic panel-data command `xtdpd`. Stata 11 made available the much needed `xtunitroot` command, which produces a fully comprehensive report on unit roots for panel data with just the click of a button, albeit available only for restrictive panel structures, plus it added options for complex residual-error structure in `xtmixed`.

Stata 13 packaged all multilevel mixed-effects models under their own domain, the [ME] class of methods, and it substantially expanded capabilities to a great range of discrete outcome and nonlinear methods. This marked a new era for the software, and Stata is now considered the leader in mixed, multilevel, and panel-data methods.

13.3.4   Time-series analysis

Stata 8 presented the first serious attempt of a competitive suite on contemporary time-series tools, with the release of a comprehensive package on vector autoregression (`var`), structural vector autoregression (`svar`) models, impulse-response functions (`irf`), and a complete smoothing package (`tssmooth`). This appealed greatly to the finance discipline. Stata 9 further modernized the [TS] suite with a `sarima()` model (multiplicative seasonal autoregressive integrated moving average) and `rolling` window and recursive estimation.

---

13. According to Google Scholar, as of 30 August 2014, Roodman (2009) has been cited 2,341 times.

Stata 10 expanded its understanding of frequency to millisecond data, but this failed to attract the tick-by-tick microstructure researchers simply because Stata cannot handle such voluminous data using standard retail computers.

Stata 11 and Stata 12 were heavily focused on time-series analysis upgrades. Stata 11 expanded time-series capabilities to estimating state-space models (`sspace`), diagonal vech multivariate generalized autoregressive conditional-heteroskedasticity models (`dvech`), and dynamic-factor models (`dfactor`). Stata 12 introduced the multivariate generalized autoregressive conditional-heteroskedasticity model (`mgarch`), the unobserved-components model (`ucm`), the autoregressive fractionally integrated moving-average model (`arfima`), and the latest filters for extracting business and seasonal cycles (`tsfilter`).

As suggested by figure 13.3, StataCorp's heavy investment in time-series capabilities has paid off immensely, because Stata is now considered to be the standard in finance research, a naturally time-series inclined discipline.

It is also important to recognize Christopher (Kit) Baum's and his coauthors' contribution to the time-series capabilities of Stata and the subsequent effect on gaining market share. Baum and Sperling (2000) developed the `dfgls` command on the Dickey–Fuller generalized least-squares unit-root test, which eventually became part of official Stata and the standard in the industry. Baum (2000) developed the `kpss` test and Baum and Wiggins (2000a,b) the `gphudak`, `modlpr`, and `roblpr` tests for detecting long memory in a time series.

13.3.5   Other key methods

One of the reasons Stata is often cited in top journals is because of Christopher Baum's, Mark Schaffer's, and Steven Stillman's work on instrumental-variable and GMM estimation (Baum, Schaffer, and Stillman 2003, 2007; Schaffer 2005). `ivregress` and `xtivreg` are a reflection of this work and are natural methods for financial and business research where the issue and treatment of endogeneity is of major concern.[14] In this respect, Stata 11 introduced the `gmm` command, which allows the user to fit models using the GMM. The syntax is accessible to even the novice, and the method presents an alternative to the more rigid maximum-likelihood estimation.

Financial and business data are plagued by missing values, and almost all published research makes the assumption that missing data are missing completely at random and casually proceeds with their pairwise deletion. Yet, it is rarely the case that our data are missing at random because data collection is systematic and even automated. Stata 11 and 12 together introduced the comprehensive [MI] suite for performing multiple imputation on missing values, and I am surprised to see that this method has yet to be picked up in volume by accounting, business, and finance.

---

14. The editor of a leading journal in accounting once explained to me that "endogeneity" is probably the single most contentious econometric concern cited by reviewers. The next one is standard-error correction, and Stata is again the appropriate software for addressing this issue.

The release of the [SEM] suite for structural equation modeling with Stata 12 brought back memories from my PhD years, when I attempted to learn LISREL because structural equation modeling seemed like a natural approach to modeling the highly structured accounting data and the relation with its market. Structural equation modeling is a type of thinking rather than a method by itself, and it has the benefit of elucidating multidirectional cause-and-effect relations and assumed variance structures. The key advantage of the `sem` command is the unified syntax that shows how a seemingly disconnected estimation models are, in fact, variations from a baseline. For example, `sem` can fit a structural system of multiple equations that may be recursive (triangular), diagonal (seemingly unrelated), or integrated (simultaneous systems).[15]

Last but not least, a special mention must be made to Colin Cameron and Pravid Trivedi's (2009) book on *Microeconometrics Using Stata*.[16] Every serious researcher in accounting and finance who uses Stata has a copy of this book. I am convinced that this voluminous work with the most attractive title has drawn many new users to Stata. The book covers nearly every tool necessary for financial research with the notable omission of time-series analysis, but this gap has been recently filled with Sean Becketti's (2013) book on *Introduction to Time Series Using Stata*.

13.4   Wishes and grumbles

The interface overhaul that started with Stata 8 seems to have missed tables. The software still lacks official publication-quality tables that can be easily produced and adapted. Ben Jann's (2005) `estout`, Roy Wada's (2005) `outreg2`, and Ian Watson's (2004) `tabout` are remarkable individual attempts, but they seem cumbersome in their implementation and have overwhelming documentation. Stata can learn from these developers and work towards a consolidated platform for generating tabular output, regardless the input, that could be easily exported to word processors ready for publication.

---

15. Triangular systems can also be estimated with reduced form `reg`, seemingly unrelated regression can be estimated with `sureg`, and integrated systems can also be estimated via `reg3`.
16. According to Google Scholar, as at 30 August 2014, Cameron and Trivedi (2009) has been cited 1,900 times.

The most critical issue is perhaps the limitation in working with very large data. Merely the concept of very large data does not fit well with Stata's ingenious yet restrictive design of placing data in memory.[17] As financial and business research draws links between traditional sources of data and the ever-expanding Internet-traffic data, retail-transaction data, social-networking, mobile-device data, and other fluid sources of information, StataCorp must either address this issue somehow or be clear that its product is not intended for analyzing very large datasets. Also it must be clearly explained that Stata may be less expensive than other software, but the user is faced with the additional cost of potential hardware upgrades every few years for accessing expanded memory storage.[18] Perhaps an authoritative monograph on *"How to work with very large data in Stata"* could address some of these concerns in the meantime.

Other issues that are often discussed in research cycles are the wish of a Bayesian platform and the lack of rotating 3D graphs (the perennial grumble at the Users Group meetings). These are not critical issues, because those using Bayesian methods typically use MATLAB or WinBugs and would take a lot of convincing to convert to Stata. However, Stata has made a name in its ability to package methods nicely, and a [BA] suite would certainly level the playing field with the frequentists. Also, even though we rarely see a 3D graph published in accounting, finance, or business journals, users of R, MATLAB, SAS, and even Excel like to pick on Stata for this omission—to them, this is a matter of prestige.

We also need more focused textbooks and more training courses not only on methods but also on specific research questions using Stata, such as "Predicting and managing bankruptcy risk using Stata", "Creditworthiness analytics using Stata", "Integrating business data sources in Stata", "Detecting earnings management using Stata", "Sourcing and mining the Electronic Data Gathering, Analysis, and Retrieval archive in Stata", etc.[19] This is not a StataCorp job per se, but the company could provide the training platform and make an open call for authors to take up this opportunity.

Finally, I would suggest that Stata sends people on the field and talk to business schools about what more Stata could offer to be the software of choice. For example, the Wharton School of the University of Pennsylvania has paired with SAS in developing and maintaining the Wharton Research Data Services, which is now considered to be the standard in financial data access globally. Such strategy of course needs resources, but the payoff would be considerable and persist over time.

---

17. Here, I define "very large data" as the data file that exceeds the available RAM. The formal definition of "big data" is simply noncompatible with Stata's architecture.

18. During the last few years, Stata has been marketed as a software for data up to 6GB of RAM (given that standard notebooks could take up to 8GB). The new generation of computer hardware can take up to 16GB of RAM, so Stata can now handle double as much.

19. It is an archive of all company forms that are required by law to be filed to the U.S. Securities and Exchange Commission. The database is freely available to the public via the Internet (http://www.sec.gov/edgar.shtml), and it is a key source of publicly available data for research purposes.

13.5   Conclusion

Stata has done exceedingly well during the last decade. It is now the most preferred software overall in finance, and the second in accounting (by a small margin) and in business (by a large margin), and has managed to overtake one of its key competitors, SAS, in scholarly research usage.

Importantly, Stata is the software of choice for high-impact research in all areas. Given that software are cited in journals (particularly in high-impact journals) due to the description or promotion of advanced methods, it can also be said that Stata has accelerated the application of state-of-the-art methods and the advancement of knowledge.

Stata's strategy of investing heavily in panel data and mixed and multilevel methods, as well as in the modernization of the time-series analysis suite and key tools in standard-error correction, and instrumental-variable treatment, including the GMM kit, has paid off immensely. Beginning with version 8, Stata has also managed to eradicate the false perception that it is not useful for data management (spread mostly by users of SAS, Excel, SPSS). Stata is now the choice software for data management, and it is also a key player in data visualization.

Stata's most critical challenge remains the issue of big data. Financial and business data are becoming bigger, multidimensional, and increasingly linked. For example, microstructure research simply does not consider Stata. So the question is whether Stata will lose relevance in other areas that require access to increasingly sizable data. Stata must also address, urgently, the issue of publication-quality tables.

Lastly, the data on software citations suggest that Stata is a leader because of its accessible state-of-the-art methods. Therefore, a recommendation to StataCorp would be to push the boundaries to other areas that are not widely applied simply because they are inaccessible. The obvious is Bayesian methods. Colin Cameron also asks for semiparametric and maximum simulated likelihood methods. StataCorp could even create a small Stata X division (as in Google X) that would take in experimental user-written work and their own work-in-progress and confidentially communicate beta software components to key users for trial—this would expedite their development into mature salable upgrades.

References

Baum, C. F. 2000. sts15: Tests for stationarity of a time series. *Stata Technical Bulletin* 57: 36–39. Reprinted in *Stata Technical Bulletin Reprints*, vol. 10, pp. 356–360. College Station, TX: Stata Press.

Baum, C. F., M. E. Schaffer, and S. Stillman. 2003. Instrumental variables and GMM: Estimation and testing. *Stata Journal* 3: 1–31.

———. 2007. Enhanced routines for instrumental variables/generalized method of moments estimation and testing. *Stata Journal* 7: 465–506.

Baum, C. F., and R. Sperling. 2000. sts15_1: Tests for stationarity of a time series. *Stata Technical Bulletin* 58: 35–36. Reprinted in *Stata Technical Bulletin Reprints*, vol. 10, pp. 360–362. College Station, TX: Stata Press.

Baum, C. F., and V. Wiggins. 2000a. dm81: Utility for time-series data. *Stata Technical Bulletin* 57: 2–4. Reprinted in *Stata Technical Bulletin Reprints*, vol. 10, pp. 29–30. College Station, TX: Stata Press.

———. 2000b. sts16: Tests for long memory in a time series. *Stata Technical Bulletin* 57: 39–44. Reprinted in *Stata Technical Bulletin Reprints*, vol. 10, pp. 362–368. College Station, TX: Stata Press.

Becketti, S. 2013. *Introduction to Time Series Using Stata*. College Station, TX: Stata Press.

Cameron, A. C., and P. K. Trivedi. 2009. *Microeconometrics Using Stata*. College Station, TX: Stata Press.

Jann, B. 2005. Making regression tables from stored estimates. *Stata Journal* 5: 288–308.

Roodman, D. M. 2009. How to do xtabond2: An introduction to difference and system GMM in Stata. *Stata Journal* 9: 86–136.

Schaffer, M. E. 2005. xtivreg2: Stata module to perform extended IV/2SLS, GMM and AC/HAC, LIML and *k*-class regression for panel data models. Statistical Software Components S456501, Department of Economics, Boston College. http://ideas.repec.org/c/boc/bocode/s456501.html.

Wada, R. 2005. outreg2: Stata module to arrange regression outputs into an illustrative table. Statistical Software Components S456416, Department of Economics, Boston College. http://ideas.repec.org/c/boc/bocode/s456416.html.

Watson, I. 2004. tabout: Stata module to export publication quality cross-tabulations. Statistical Software Components S447101, Department of Economics, Boston College. http://ideas.repec.org/c/boc/bocode/s447101.html.

13.6 About the author

Demetris Christodoulou is a senior lecturer at the University of Sydney Business School and the general convenor of the research network *Methodological and Empirical Advances in Financial Analysis*. He is the architect of the MEAFA Professional Development Workshops on Quant Analysis Using Stata with extensive consulting experience on Stata in academia, industry, and government.

A. Appendix. Software citations in scholarly journals in accounting, business, and finance, 1999–2013

Journal	SPSS	Stata	SAS	MATLAB	RATS	EViews	SPlus	Limdep	SJR
ASTIN Bulletin [F]	0	1	0	0	0	0	0	0	1.08
Abacus [A]	6	5	7	0	2	0	0	1	0.36
Acad of Acc & Fin Stu J [A]	16	5	13	0	1	4	0	1	0.18
Acad of Ban Stu J [F]	0	0	3	0	0	1	0	0	0.10
Acc & Bus Res [A]	8	8	1	0	0	0	1	2	0.91
Acc & Fin [A]	14	11	7	4	8	1	1	2	0.58
Acc & Mgt Inf Sys [A]	12	0	0	0	0	1	0	0	
Acc & Taxation [A]	3	0	1	1	0	1	0	0	
Acc Edu [A]	26	0	1	0	0	0	0	0	0.58
Acc Forum [A]	8	0	3	0	0	0	0	0	0.67
Acc Historians J [A]	2	0	2	0	0	0	0	0	0.00
Acc History [A]	0	0	0	0	0	0	0	0	0.30
Acc History Rev [A]	1	0	3	0	1	0	1	0	0.22
Acc Horizons [A]	2	4	10	0	2	0	0	0	1.35
Acc Persp [A]	1	2	4	0	0	0	0	0	0.16
Acc Res J [A]	6	0	0	0	0	0	0	0	0.15
Acc Rev [A]	5	38	65	1	1	1	0	1	5.00
Acc in Eur [A]	1	0	2	0	0	0	0	0	0.61
Acc, Accountability & Perf [A]	4	1	0	0	0	0	0	0	
Acc, Aud & Accountability J [A]	5	0	3	0	0	0	0	1	0.97
Acc, Bus & Fin History [A]	0	0	0	0	0	0	0	0	
Acc, Mgt & Inf Tech [A]	0	1	0	0	0	0	0	0	
Acc, Org & Soc [A]	15	5	10	0	0	0	1	1	2.09
Adv in Acc [A]	14	11	10	0	0	0	0	0	0.25
Adv in Int Acc [A]	3	0	0	0	0	0	0	0	0.00
Afr J of Acc, Aud & Fin [A]	1	0	0	0	1	1	5	0	
Afr J of Acc,Eco,Fin & Ban Res [A]	9	0	1	0	0	0	0	1	0.00
Afr Rev of Mon Fin & Ban [F]	3	2	3	3	1	3	0	1	0.11
Afro-Asi J of Fin & Acc [F]	0	0	0	0	0	0	0	0	
Agenda: J of Pol Analysis & Ref [F]	0	0	0	0	0	1	0	0	
AgrBus [B]	13	0	14	2	2	5	1	10	0.47
Ame Bankers Asso Ban J [F]	0	0	0	0	0	0	0	0	
Ame Bankruptcy Insti J [F]	0	2	0	0	0	0	0	0	0.37
Ame Bankruptcy Law J [F]	5	10	0	0	0	0	0	0	0.45
Ame Bus Law J [B]	0	0	0	0	0	0	0	0	
Ame J of Bus [B]	6	2	0	0	0	0	0	0	
Ame J of Bus Edu [B]	2	3	3	1	0	2	0	19	
Ame J of Bus Res [B]	3	1	1	0	0	0	0	0	
Ame J of Fin & Acc [F]	0	0	0	0	0	0	0	0	
Annals of Actuarial Sci [F]	0	0	2	1	0	0	0	0	0.39
Annals of Eco & Fin [F]	0	0	0	0	0	0	0	0	0.79
Annals of Fin [F]	0	0	0	3	0	0	0	0	3.71
Annamalai Int J of Bus Stu&Res [B]	9	0	0	0	0	0	0	0	
Annual Rev of Fin Eco [F]	0	0	0	1	0	0	0	0	0.28
App Fin Eco [F]	6	23	10	31	20	38	7	4	0.00
App Fin Eco Letters [F]	6	63	25	26	24	30	8	23	1.00
App Math Fin [F]	0	0	0	21	0	1	0	0	0.51
App Stoch Models in Bus & Ind [B]	8	6	27	32	3	4	15	1	
Arab Eco & Bus J [B]	1	0	0	0	0	0	0	0	0.28
Asi Bus & Mgt [B]	9	11	2	0	0	0	0	0	
Asi J of Bus Ethics [B]	0	1	0	0	0	0	0	0	0.38
Asi Pac Bus Rev [B]	17	3	3	0	1	3	0	0	
Asi Pac J of Eco & Bus [B]	0	3	1	0	0	0	0	0	0.17
Asi Pac J of Fin & Ban Res [F]	2	1	0	2	0	1	0	0	0.38
Asi Rev of Acc [A]	11	2	0	0	0	3	0	17	0.20
Asi-Pac Bus Rev [B]	0	3	3	0	1	1	0	0	0.22
Asi-Pac Fin markets [F]	0	1	4	5	5	0	0	0	0.24
Asi-Pac J of Acc & Eco [A]	1	0	0	0	0	0	0	0	0.32
Asi-Pac J of Bus Admin [B]	6	1	1	0	0	1	1	0	1.59
Asi-Pac J of Fin Stu [F]	0	7	0	0	6	1	0	0	0.49
Aud: A J of Pract & Theory [A]	5	13	15	0	0	0	0	0	
Australian Acc Rev [A]	4	0	2	0	0	4	0	0	
Balance Sheet [A]	0	0	0	0	0	0	0	0	
Ban & Fin Law Rev [F]	0	0	0	0	0	0	0	0	
Ban & Fin Letters [F]	2	0	0	0	0	2	0	0	
Ban & Fin Rev [F]	0	0	1	0	0	0	0	0	0.15
Banca Naz del Lavoro Quar Rev [F]	0	1	0	0	3	2	0	0	
Bank of England Quar Bulletin [F]	0	0	0	0	0	1	0	0	
Banks & Bank Sys [F]	3	0	0	0	0	0	0	0	
Banks in Ins Report [F]	0	0	0	0	0	0	0	0	
Barclays Eco Rev [F]	0	0	0	0	0	0	0	0	
Baylor Bus Rev [B]	0	1	4	0	0	0	0	0	
Beh Res in Acc [A]	12	1	11	0	0	0	0	0	0.48
Benefits Law J [A]	0	0	0	0	0	0	0	0	
Borsa Istanbul Rev [B]	0	0	0	0	0	1	1	2	0.63
British Acc Rev [A]	20	8	3	0	0	0	0	0	
British Actuarial J [F]	0	0	0	1	0	0	0	0	
Bus & Eco History [B]	0	0	0	0	0	0	0	0	
Bus & Eco Res [B]	10	2	1	0	0	0	0	0	

Continued on next page

Journal	SPSS	Stata	SAS	MATLAB	RATS	EViews	SPlus	Limdep	SJR
Bus & Politics [B]	0	14	0	1	0	0	1	0	0.51
Bus & Prof Ethics J [B]	2	0	1	0	0	0	0	0	.
Bus Comm Quar [B]	0	0	0	0	0	0	0	0	0.27
Bus Eco [B]	2	0	0	2	2	7	0	0	0.15
Bus Edu & Accreditation [B]	6	1	1	0	0	0	0	0	.
Bus Ethics Quar [B]	0	0	0	0	0	0	0	0	2.30
Bus Ethics: A Eur Rev [B]	15	0	3	0	4	0	1	0	.
Bus Exc [B]	0	0	0	0	0	0	0	0	.
Bus Forum [B]	1	0	0	0	0	0	0	0	.
Bus History [B]	0	0	1	0	0	0	0	0	0.32
Bus History Rev [B]	0	0	0	0	0	0	0	0	0.46
Bus Horizons [B]	1	1	8	0	5	0	1	0	1.52
Bus Inf Rev [B]	5	1	7	0	0	0	0	0	0.25
Bus Persp [B]	0	0	1	0	0	0	0	0	.
Bus Renaissance Quar [B]	6	0	1	0	0	0	0	0	.
Bus Res Quar [B]	0	0	0	0	0	0	0	0	.
Bus Strategy & the Envi [B]	26	12	7	1	3	0	1	0	1.23
Bus Strategy Rev [B]	0	1	3	0	4	0	0	0	0.11
Bus&Soc [B]	2	1	2	0	0	0	1	0	1.04
Bus&Soc Rev [B]	4	0	2	0	5	0	0	0	0.23
Bus: Theory & Pract [B]	0	0	0	0	0	0	0	0	0.21
CPA J [A]	1	0	2	0	0	0	0	0	.
Canadian Bus Eco [B]	0	0	0	0	0	0	0	0	.
Cent Bank Rev [F]	2	1	0	0	0	0	0	0	.
Chin - USA Bus Rev [B]	35	5	2	3	1	6	1	0	.
Chin Bus Rev [B]	46	6	5	2	2	6	1	2	.
Chin Fin Rev Int [F]	2	3	1	5	0	1	0	0	.
Chin J of Acc Res [A]	0	4	1	0	0	0	0	0	.
Chin J of Acc Stu [A]	0	0	0	0	0	0	0	0	.
Cont Acc Res [A]	3	28	21	0	3	0	3	1	2.54
Contab y Negocios [A]	3	1	0	0	0	0	0	0	.
Corp Fin Rev [F]	0	0	0	0	0	0	1	0	.
Corp Gov [A]	14	24	9	0	5	1	0	2	1.16
Crit Fin Rev [F]	0	0	0	1	0	0	0	0	.
Crit Persp on Acc [A]	8	0	4	0	0	0	0	0	0.92
Crit Persp on Int Bus [B]	2	0	0	0	0	0	0	0	0.31
Cuad de Eco y Dir de la Emp [B]	32	11	7	0	0	0	0	1	.
Current Iss in Aud [A]	0	0	0	0	0	0	0	0	0.32
Current Iss in Eco & Fin [F]	0	0	0	0	0	0	0	0	.
Decisions in Eco & Fin [F]	0	0	0	6	0	0	0	0	0.12
Defense Counsel J [F]	0	0	0	0	0	0	0	0	0.00
Eco & Bus Rev for Cen&SE Eur [B]	16	3	1	0	0	1	0	0	.
Eco, Mgt & Fin Mark [F]	6	6	0	1	0	1	0	0	.
Edu Fin & Pol [F]	0	12	4	0	0	0	0	0	1.51
Edu,Bus&Soc:Cont Mid-Eas Iss [B]	17	0	0	0	0	0	0	0	0.24
Emrg Mark Rev [F]	0	18	4	4	3	4	2	0	0.57
Emrg Mark, Fin & Trade [F]	0	0	0	0	2	0	0	0	0.30
Enterprise & Soc [B]	0	1	0	0	0	0	0	0	.
Enterprise Risk Mgt [F]	1	0	0	0	0	0	0	0	.
Eur Acc Rev [A]	14	9	8	1	1	0	0	1	0.83
Eur Bus J [B]	0	0	0	0	0	0	0	0	.
Eur Bus Org Law Rev [B]	0	0	3	0	0	0	0	0	0.24
Eur Bus Rev [B]	22	1	3	0	0	0	0	0	0.37
Eur Fin Mgt [F]	1	8	4	3	10	0	0	1	1.39
Eur J of Fin [F]	7	3	3	6	1	1	4	0	0.43
Eur J of Fin & Ban Res [F]	2	0	0	0	0	0	0	0	.
Eur J of Risk Reg [F]	0	0	0	0	0	0	0	0	0.26
EuroMed J of Bus [B]	22	0	3	2	0	0	0	0	0.00
FDCC Quar [F]	0	0	0	0	0	0	0	0	.
Family Bus Rev [B]	5	1	3	0	0	0	0	0	2.69
Fin & Stoch [F]	0	0	0	5	0	0	2	0	2.00
Fin Accountability & Mgt [A]	9	2	1	0	2	0	0	0	.
Fin Analysts J [F]	0	1	2	11	1	0	2	0	1.93
Fin History [F]	0	0	0	0	0	0	0	0	.
Fin History Rev [F]	1	2	0	0	0	0	0	0	1.36
Fin Ind Persp [F]	0	0	0	0	0	0	0	0	.
Fin India [F]	5	0	4	2	1	0	2	0	.
Fin Mark & Port Mgt [F]	5	1	0	1	1	2	1	0	0.64
Fin Mark, Insti & Instr [F]	0	6	2	1	5	1	0	0	0.38
Fin Mgt [F]	0	15	2	1	6	0	0	0	1.95
Fin Res Letters [F]	0	1	2	8	1	0	2	0	0.47
Fin Rev [F]	4	4	7	1	5	3	0	0	0.93
Fin Services Rev [F]	0	0	2	0	0	0	0	1	.
Fin Theory & Pract [F]	2	7	0	0	0	3	0	0	.
Fin a Uver [F]	0	7	0	6	0	3	0	0	0.25
FinanzArchiv [F]	0	6	1	1	0	0	0	0	0.30
Foundations & Trends in Acc [A]	0	0	0	0	0	0	0	0	0.46
Foundations & Trends in Fin [F]	0	0	0	0	0	0	0	0	1.19
Frontiers of Bus Res in Chin [B]	32	5	1	1	0	0	0	0	0.19
GITAM Rev of Int Bus [B]	0	0	0	0	0	0	0	0	.
Gadjah Mada Int J of Bus [B]	0	0	0	0	0	0	0	0	.
Geneva Papers on Risk & Ins [F]	2	12	6	3	1	1	0	0	0.10
Geneva Risk & Ins Rev [F]	0	0	1	2	0	0	0	0	0.62
German Bus Rev [B]	0	0	0	1	0	0	0	0	0.34
Glo Bus & Eco Rev [B]	0	0	3	0	0	1	0	0	0.13
Glo Bus Rev [B]	20	3	9	2	0	1	0	2	0.21

Continued on next page

Journal	SPSS	Stata	SAS	MATLAB	RATS	EViews	SPlus	Limdep	SJR
Glo Fin J [F]	1	1	5	2	6	3	0	1	0.38
Glo J of Bus Res [B]	8	0	1	0	0	0	0	0	.
Glo J of Fin & Ban Iss [F]	1	1	1	1	0	0	0	0	.
Glo J of Int Bus Res [B]	1	0	0	0	1	0	0	0	.
Glo Persp on Acc Edu [A]	0	0	0	0	0	0	0	0	.
Harvard Bus Rev [B]	1	0	3	0	0	0	0	0	0.42
Housing Fin Int [F]	1	0	0	0	0	0	0	0	.
IAMURE Int J of Bus & Mgt [B]	4	0	1	0	0	0	0	0	.
ISM J of Int Bus [B]	2	0	0	0	0	0	0	0	.
ISRA Int J of Isl Fin [F]	0	0	0	0	0	0	0	0	.
IUP J of Acc Res & Aud Pract [A]	7	1	0	0	0	0	0	0	.
IUP J of App Fin [F]	9	2	0	2	2	10	0	0	.
IUP J of Bank Mgt [F]	17	0	2	0	0	1	0	0	.
IUP J of Beh Fin [F]	5	1	0	0	0	0	0	0	.
IUP J of Fin Eco [F]	1	0	0	0	1	2	0	0	.
IUP J of Fin Risk Mgt [F]	1	0	0	3	0	0	0	0	.
Illinois Bus Rev [B]	0	0	0	0	0	0	0	0	.
InFin [F]	0	0	0	0	0	0	0	0	.
Indiana Bus Rev [B]	0	0	0	0	0	0	0	0	0.31
Inf Sys & e-Bus Mgt [B]	9	2	3	2	0	0	1	0	1.18
Ins: Math & Eco [F]	3	2	15	65	2	2	18	1	0.94
Int Bus Rev [B]	33	15	15	0	1	2	5	5	1.05
Int Fin [F]	0	5	0	3	6	2	0	1	0.39
Int Food & AgrBus Mgt Rev [B]	0	7	0	0	0	1	1	3	0.13
Int Insolvency Rev [F]	1	1	2	0	1	0	0	0	0.39
Int J of Acc [A]	5	8	7	0	0	1	1	1	.
Int J of Acc & Fin [A]	0	0	1	0	0	0	0	0	.
Int J of Acc & Fin Reporting [A]	7	5	2	2	0	3	0	0	0.27
Int J of Acc & Inf Mgt [A]	6	3	2	0	0	0	0	0	0.87
Int J of Acc Inf Sys [A]	14	2	6	0	0	0	0	0	0.15
Int J of Acc, Aud & Perf Eval [A]	0	0	1	0	0	0	0	0	.
Int J of App Eco & Fin [F]	0	0	0	1	0	0	0	0	0.39
Int J of Aud [A]	17	4	7	0	1	0	0	0	0.11
Int J of Ban, Acc & Fin [F]	0	0	2	0	0	1	0	0	.
Int J of Beh Acc & Fin [A]	0	0	0	0	0	0	0	0	.
Int J of Bonds & Deriv [F]	0	0	0	0	0	0	1	0	0.12
Int J of Bus [B]	1	1	3	1	1	0	0	0	.
Int J of Bus & Commerce [B]	2	0	0	0	0	0	0	1	.
Int J of Bus & Eco [B]	5	2	2	3	2	0	0	0	.
Int J of Bus & Emrg Mark [B]	0	0	1	0	0	2	0	0	.
Int J of Bus & Fin Res [B]	3	2	1	1	0	8	1	0	.
Int J of Bus & Glo [B]	0	0	2	0	0	1	0	0	0.13
Int J of Bus & Inf [B]	9	0	0	0	0	0	0	0	.
Int J of Bus & Mgt Sci [B]	16	0	0	0	0	0	0	0	0.10
Int J of Bus & Sys Res [B]	0	0	0	1	0	0	0	0	.
Int J of Bus Comp & Growth [B]	0	0	2	0	0	1	0	0	.
Int J of Bus Cont & Risk Mgt [B]	0	0	6	2	0	0	0	0	.
Int J of Bus Envi [B]	0	0	3	0	0	0	0	0	0.31
Int J of Bus Exc [B]	0	0	0	0	0	0	0	0	.
Int J of Bus For & Markg Intel [B]	0	0	0	0	0	0	0	0	0.19
Int J of Bus Gov & Ethics [B]	0	0	2	0	0	2	0	0	0.32
Int J of Bus Inf Sys [B]	2	0	4	2	0	3	0	0	0.25
Int J of Bus Inno & Res [B]	0	0	0	0	0	3	0	0	0.14
Int J of Bus Intel & Data Mining [B]	0	0	2	0	0	1	0	0	.
Int J of Bus Mgt & Eco Res [B]	11	1	2	1	0	1	0	0	0.13
Int J of Bus Perf Mgt [B]	0	0	0	0	0	2	0	0	0.24
Int J of Bus Proc Integ & Mgt [B]	0	0	0	0	0	3	0	0	.
Int J of Bus Stu [B]	4	1	0	0	0	3	0	1	0.12
Int J of Bus&Soc [B]	19	2	3	2	1	3	0	0	.
Int J of Crit Acc [A]	0	1	0	0	0	1	0	0	0.33
Int J of Disclosure & Gov [A]	3	0	2	0	0	1	0	0	.
Int J of Eco & Acc [A]	1	0	0	0	0	1	0	0	.
Int J of Eco & Bus Modeling [B]	2	0	1	2	0	0	0	0	.
Int J of Eco & Bus Res [B]	0	0	1	0	0	0	0	0	.
Int J of Eco, Mgt & Acc [A]	1	2	1	0	1	0	0	0	.
Int J of Electronic Ban [F]	0	0	0	0	0	4	0	0	.
Int J of Electronic Bus [B]	0	0	0	0	0	2	0	0	.
Int J of Electronic Fin [F]	0	0	1	0	0	0	0	0	0.46
Int J of Fin & Eco [F]	3	8	0	8	17	6	0	0	.
Int J of Fin Engi & Risk Mgt [F]	0	0	0	0	0	0	0	0	.
Int J of Fin Mark & Deriv [F]	0	0	3	0	0	0	0	0	.
Int J of Fin Mgt [F]	12	0	0	3	0	4	0	0	.
Int J of Fin Services Mgt [F]	0	0	0	0	0	3	0	0	.
Int J of For [F]	15	12	36	33	15	15	10	4	1.61
Int J of Glo & Small Bus [B]	0	0	0	0	0	1	0	0	0.16
Int J of Glo Bus [B]	3	0	0	0	0	0	0	0	.
Int J of Health Care Fin & Eco [F]	6	58	12	0	2	0	0	12	0.50
Int J of Indian Cult & Bus Mgt [B]	3	0	0	1	1	0	0	0	.
Int J of Isl & Mid-Eas Fin&Mgt [F]	11	0	1	0	1	1	1	0	0.27
Int J of Managerial Fin [F]	0	2	1	4	0	0	0	1	.
Int J of Mgt Scis & Bus Res [B]	4	0	0	0	0	0	0	0	0.18
Int J of Monetary Eco & Fin [F]	0	0	0	0	0	0	0	0	.
Int J of Port Analysis & Mgt [F]	0	0	0	0	0	0	0	0	0.27
Int J of Sport Fin [F]	0	12	1	0	0	0	1	0	0.51
Int J of Strategic Property Mgt [F]	12	0	1	3	0	4	1	0	0.74
Int J of Theoretical & App Fin [F]	0	0	0	0	0	0	0	0	.

Continued on next page

Journal	SPSS	Stata	SAS	MATLAB	RATS	EViews	SPlus	Limdep	SJR
Int J of e-Bus Res [B]	5	0	2	0	0	0	0	0	0.15
Int J of the Eco of Bus [B]	0	0	0	0	0	0	0	0	0.54
Int Rev of App Fin Iss & Eco [F]	1	0	1	1	1	3	0	0	.
Int Rev of Eco & Fin [F]	0	12	7	7	10	3	1	2	0.68
Int Rev of Fin [F]	0	4	2	1	1	0	0	0	0.85
Int Rev of Fin Analysis [A]	1	12	8	8	6	2	5	2	0.35
Int Small Bus J [B]	28	8	3	0	0	0	0	0	0.98
Int Tax & Pub Fin [F]	0	11	2	4	1	1	0	1	1.07
Intel Sys in Acc, Fin&Mgt [A]	6	0	5	12	3	1	2	0	.
Invest Mgt & Fin Innos [F]	1	0	0	0	0	1	0	0	0.19
Iss in Acc Edu [A]	1	1	4	0	0	0	0	0	0.37
J - A&NZ Insti of Ins & Fin [F]	0	0	0	0	0	0	0	0	.
J for Glo Bus Advancement [B]	1	0	0	0	0	0	0	0	0.12
J for Int Bus & Ent Dev [B]	1	0	2	0	0	0	0	0	.
J of Acc & Aud [A]	3	0	0	0	0	0	0	0	.
J of Acc & Eco [A]	0	27	22	0	3	1	1	1	7.29
J of Acc & Fin [A]	0	1	2	0	0	1	1	0	.
J of Acc & Fin Res [A]	2	0	4	0	0	0	0	1	.
J of Acc & Org Change [A]	6	0	0	0	0	0	0	0	0.13
J of Acc & Pub Pol [A]	4	12	16	0	0	0	0	2	1.09
J of Acc Edu [A]	5	0	3	0	0	0	0	2	0.34
J of Acc Literature [A]	0	0	0	0	0	0	0	0	.
J of Acc Res [A]	0	29	18	0	13	1	0	0	5.15
J of Acc in Emrg Eco [A]	5	0	0	0	0	0	0	0	.
J of Acc, Aud & Fin [A]	1	4	16	1	0	0	0	0	0.64
J of Acc, Fin&Mgt Strategy [A]	7	0	2	0	0	1	0	0	.
J of Accountancy [A]	0	0	2	0	0	0	0	0	.
J of Advanced Stu in Fin [F]	0	0	0	0	0	1	0	0	.
J of AgrBus in Dev & Emrg Eco [B]	1	0	0	0	0	1	0	0	.
J of Alternative Invest [F]	0	2	0	0	0	0	4	0	0.92
J of App Acc Res [A]	6	1	0	0	0	0	0	0	0.17
J of App Bus & Eco [B]	16	4	2	0	0	4	0	0	.
J of App Bus Res [B]	24	4	7	2	1	2	1	0	0.16
J of App Corp Fin [F]	0	1	0	1	12	0	0	0	.
J of App Fin [F]	0	1	5	0	0	0	0	0	.
J of App Mgt Acc Res [A]	11	0	1	0	0	0	0	1	.
J of App Res for Bus Instruction [B]	0	0	0	0	0	0	0	0	.
J of Asi Bus Strategy [B]	14	1	0	0	0	0	0	0	.
J of Asi Bus Stu [B]	2	0	0	0	1	0	0	1	.
J of Asi-Pac Bus [B]	8	0	2	0	0	1	0	2	0.15
J of Asset Mgt [F]	2	1	4	15	1	1	3	0	0.19
J of Ban & Fin [F]	2	74	28	52	23	11	7	7	1.42
J of Ban Reg [F]	0	2	0	0	0	0	0	2	0.17
J of Beh & Exper Fin [F]	0	0	0	0	0	0	0	0	.
J of Beh Fin [F]	7	2	2	1	1	0	2	0	0.61
J of Bus [B]	0	1	3	4	0	0	0	1	0.00
J of Bus & Acc [B]	0	0	0	0	0	0	0	0	.
J of Bus & Behavior Scis [B]	1	1	0	0	0	0	0	0	.
J of Bus & Eco [B]	3	0	0	1	0	0	0	0	.
J of Bus & Eco Res [B]	10	1	2	2	0	0	0	0	.
J of Bus & Eco Statistics [B]	0	19	14	33	7	1	6	0	5.29
J of Bus & Eco Stu [B]	5	2	2	0	1	1	0	1	.
J of Bus & Edual Leadership [B]	2	0	0	0	0	0	0	0	.
J of Bus & Fin Librarianship [F]	8	3	6	1	1	1	0	0	0.65
J of Bus & Psychology [B]	34	0	20	0	0	0	1	0	1.81
J of Bus & Technical Comm [B]	3	0	2	0	0	0	0	0	0.42
J of Bus Admin Res [B]	4	0	1	0	0	0	0	0	.
J of Bus Cases & App [B]	0	0	0	0	0	0	0	0	.
J of Bus Comm [B]	1	0	0	0	0	0	0	0	0.79
J of Bus Diversity [B]	4	1	1	0	0	0	0	0	.
J of Bus Eco & Mgt [B]	21	4	1	7	2	4	1	0	0.51
J of Bus Ethics [B]	163	29	56	2	0	3	3	2	0.96
J of Bus Ethics Edu [B]	1	0	0	0	0	0	0	0	.
J of Bus Exc [B]	2	0	0	0	0	0	0	0	.
J of Bus For [B]	9	0	29	0	0	0	1	0	.
J of Bus Logistics [B]	21	0	5	0	0	0	0	0	1.87
J of Bus Res [B]	120	34	62	2	7	2	0	4	1.22
J of Bus Strategies [B]	12	0	1	0	0	0	0	0	.
J of Bus Strategy [B]	0	0	1	0	0	0	0	0	0.40
J of Bus Theory & Pract [B]	0	0	0	0	0	0	0	0	.
J of Bus Venturing [B]	24	42	15	1	3	0	3	5	4.36
J of Bus, Fin & Acc [A]	10	30	17	3	9	1	3	7	1.01
J of Chin Eco & Bus Stu [B]	0	3	0	1	0	3	0	0	0.29
J of Commerce & Acc Res [A]	6	1	0	1	0	1	0	0	.
J of Cont Acc & Eco [A]	2	0	2	0	0	0	0	0	0.39
J of Corp Acc & Fin [A]	1	1	28	0	21	1	0	0	.
J of Corp Citizenship [F]	2	0	0	0	0	0	0	0	.
J of Corp Fin [F]	0	41	9	5	4	1	1	4	1.84
J of Corp Real Estate [F]	7	0	0	0	0	0	0	0	0.00
J of Credit Risk [F]	0	0	6	4	0	0	2	0	.
J of Deriv [F]	0	0	1	1	0	0	0	0	0.43
J of Deriv & Hedge Funds [F]	1	1	0	7	2	2	5	0	0.21
J of EU Res in Bus [B]	0	0	0	0	0	0	0	0	.
J of East Eur Res in Bus & Eco [B]	3	0	0	0	0	1	0	0	.
J of East-West Bus [B]	8	1	1	0	0	1	0	0	0.17
J of Eco & Bus [B]	1	12	8	7	9	1	2	3	0.36

Continued on next page

Journal	SPSS	Stata	SAS	MATLAB	RATS	EViews	SPlus	Limdep	SJR
J of Eco & Fin [F]	3	3	6	1	3	1	1	0	0.24
J of Eco, Fin & Admin Sci [F]	2	0	0	0	0	0	0	0	.
J of Edu Fin [F]	8	5	16	1	0	0	0	1	0.36
J of Edu for Bus [B]	32	3	10	0	0	0	0	0	.
J of Empi Fin [F]	0	7	4	21	5	3	3	2	1.29
J of Emrg Market Fin [F]	2	1	1	2	3	5	0	0	0.22
J of Emrg Tech in Acc [A]	3	0	0	0	0	0	0	0	0.35
J of Eur Bus [B]	0	0	0	0	0	0	0	0	.
J of Eur Real Estate Res [F]	2	0	3	0	1	0	0	0	0.33
J of Family Bus Strategy [B]	11	2	4	0	0	0	0	0	0.96
J of Fin [F]	1	38	14	5	41	1	3	1	18.44
J of Fin & Eco Pract [F]	1	0	0	0	0	2	0	0	.
J of Fin & Quant Analysis [F]	0	19	2	2	0	0	0	0	4.74
J of Fin Counsel & Plann [F]	16	3	23	0	0	0	0	0	0.33
J of Fin Crime [F]	1	2	2	0	0	0	0	0	0.21
J of Fin Eco [F]	0	58	16	12	11	0	6	1	11.53
J of Fin Eco Pol [F]	0	3	1	2	0	0	0	0	.
J of Fin Econometrics [F]	1	1	3	23	1	2	9	0	1.46
J of Fin Intermediation [F]	0	9	5	1	4	0	1	0	4.68
J of Fin Mark [F]	0	3	5	1	1	1	1	1	2.78
J of Fin Mgt & Analysis [F]	5	0	2	2	0	0	0	0	0.00
J of Fin Reg & Compli [F]	1	1	0	1	0	1	0	1	.
J of Fin Reporting & Acc [A]	6	0	0	0	0	2	0	0	.
J of Fin Res [F]	1	4	5	0	3	0	0	0	0.61
J of Fin Service Profs [F]	1	0	0	0	0	0	0	0	.
J of Fin Services Markg [F]	33	1	7	0	0	0	2	1	0.24
J of Fin Services Res [F]	0	11	3	2	1	2	2	5	1.20
J of Fin Stability [F]	1	5	1	1	4	0	1	0	1.66
J of Fin Stu&Res [F]	1	0	0	0	0	2	0	0	.
J of Fin, Acc & Mgt [A]	1	0	0	0	0	0	0	0	.
J of Fixed Income [F]	0	4	7	2	1	0	0	0	0.34
J of Foodservice Bus Res [B]	81	4	12	1	0	0	0	0	0.26
J of For [F]	1	1	5	5	9	3	1	1	0.91
J of Futures Mark [F]	0	1	9	23	11	3	3	0	0.70
J of Glo Bus & Tech [B]	20	0	2	0	0	0	0	0	.
J of Glo Bus Admin [B]	2	0	0	1	0	1	0	0	.
J of Glo Bus Iss [B]	10	2	3	0	0	0	0	0	.
J of Glo Bus Mgt [B]	25	3	0	0	0	0	0	0	.
J of Government Fin Mgt [A]	3	0	2	0	0	0	1	0	.
J of Hospitality Fin Mgt [F]	14	1	5	0	0	0	0	0	.
J of Indian Bus Res [B]	10	0	0	1	0	0	0	0	.
J of Indonesian Economy & Bus [B]	6	1	0	0	0	0	0	0	.
J of Inf Sys [A]	4	2	8	0	0	0	0	0	1.16
J of Inno & Bus Best Pract [B]	2	0	0	0	0	0	0	0	.
J of Ins Iss [F]	4	2	12	4	2	0	0	2	.
J of Ins Reg [F]	0	6	3	0	0	0	0	0	.
J of Int Acc Res [A]	1	2	1	0	0	0	0	1	0.63
J of Int Acc, Aud & Taxation [A]	8	3	5	0	0	0	0	1	0.59
J of Int Bus Edu [B]	0	0	0	0	0	0	0	0	.
J of Int Bus Ethics [B]	1	0	0	0	0	0	0	0	.
J of Int Bus Res [B]	9	2	5	1	0	2	0	0	.
J of Int Bus Stu [B]	26	55	32	2	0	1	3	2	4.84
J of Int Edu in Bus [B]	2	0	0	0	0	0	0	0	.
J of Int Fin Mark, Insti & Mon [F]	0	13	2	7	10	6	1	3	0.49
J of Int Fin Mgt & Acc [A]	2	3	5	0	1	0	0	0	0.54
J of Int Mon & Fin [F]	1	28	4	19	17	12	4	5	1.24
J of Intellectual Capital [A]	22	2	3	0	0	2	0	0	0.79
J of Invest Compli [F]	0	0	0	0	0	0	0	0	.
J of Isl Acc & Bus Res [B]	4	0	0	0	0	0	0	0	.
J of Law, Fin & Acc [F]	0	0	0	0	0	0	0	0	.
J of Mgt & Gov [F]	13	8	6	2	0	0	0	2	.
J of Mgt Acc Res [A]	11	0	0	0	0	0	0	0	0.22
J of Mon Laundering Control [F]	1	0	0	0	0	0	0	0	.
J of Mon, Credit & Ban [F]	0	24	3	21	9	4	5	0	2.13
J of Multinational Fin Mgt [F]	1	3	4	0	3	1	1	0	0.21
J of Pension Eco & Fin [F]	0	8	2	4	0	0	0	0	0.29
J of Perf Mgt [F]	3	0	1	0	0	0	0	0	.
J of Personal Fin [F]	3	0	2	0	0	0	0	0	.
J of Pharma Fin, Eco & Pol [F]	1	0	0	0	0	0	0	0	0.00
J of Port Mgt [F]	0	2	5	10	0	0	5	0	0.77
J of Private Equity [F]	5	3	2	0	0	0	0	0	0.14
J of Property Invest & Fin [F]	13	1	0	3	1	3	1	3	0.47
J of Pub Budg, Acc & Fin Mgt [A]	12	5	3	0	0	2	0	0	0.22
J of Real Estate Fin & Eco [F]	3	18	20	6	10	2	4	1	1.31
J of Real Estate Literature [F]	2	0	2	1	0	1	0	0	0.44
J of Real Estate Port Mgt [F]	6	1	5	5	2	0	2	1	0.21
J of Res in Bus Edu [B]	13	0	0	0	0	0	0	0	.
J of Revenue & Pricing Mgt [A]	3	0	14	7	0	0	1	0	0.46
J of Risk [F]	0	1	1	12	0	1	2	0	.
J of Risk & Ins [F]	1	16	20	5	15	6	5	4	0.92
J of Risk & Uncertainty [F]	0	19	1	5	0	0	1	3	2.34
J of Risk Fin [F]	6	5	5	6	0	2	2	1	.
J of Risk Model Validation [F]	0	0	9	3	0	0	0	0	.
J of Risk Res [F]	27	1	7	2	0	0	0	0	0.64
J of Small Bus & Enterprise Dev [B]	53	1	3	0	0	1	0	0	0.29
J of Small Bus & Entrep [B]	20	7	1	0	2	1	0	0	.

Continued on next page

Journal	SPSS	Stata	SAS	MATLAB	RATS	EViews	SPlus	Limdep	SJR
J of Small Bus Mgt [B]	31	6	7	0	0	1	0	1	1.12
J of Structured Fin [F]	0	1	0	1	0	0	0	0	.
J of Sustainable Fin & Invest [F]	1	0	0	0	0	1	0	0	.
J of Teach in Int Bus [B]	14	0	1	0	0	0	0	0	0.22
J of Theoretical Acc Res [A]	4	0	2	0	1	0	0	0	.
J of Trust Res [F]	0	1	0	0	0	0	0	0	.
J of Wealth Mgt [F]	0	0	2	2	1	3	2	0	0.13
J of World Bus [B]	26	17	9	0	0	1	0	2	1.76
J of World Energy Law & Bus [B]	0	0	0	0	0	0	0	0	0.32
J of the Ame Taxation Asso [A]	0	6	8	0	0	2	0	0	0.94
JASSA [F]	6	2	0	0	0	1	0	0	0.10
La Revue Gestion et Org [B]	13	3	0	1	0	0	1	0	.
Lat Ame Bus Rev [B]	17	3	4	1	1	0	0	0	0.13
MENA J of Bus Case Stu [B]	0	0	0	0	0	0	0	0	.
Managerial Aud J [A]	50	8	8	1	0	0	0	0	0.29
Managerial Fin [F]	14	9	21	5	3	5	3	1	.
Math & Fin Eco [F]	0	0	0	3	0	0	0	0	0.70
Math Fin [F]	0	0	3	24	0	0	2	0	1.91
Measuring Bus Exc [B]	20	0	9	0	0	0	0	0	0.44
Meditari Accountancy Res [A]	11	1	6	0	0	1	0	0	.
Mgt Acc Res [A]	10	1	6	0	0	0	0	0	0.85
Mgt Sci & Fin Engi [F]	1	0	1	2	0	0	0	0	.
Mid-Atlantic J of Bus [B]	1	0	0	0	0	0	0	0	.
Mid-Eas Bus & Eco Rev [B]	2	0	0	0	1	1	0	0	0.10
Montana Bus Quar [B]	0	0	0	0	0	0	0	0	.
Multinational Bus Rev [B]	4	6	6	0	1	0	0	1	0.11
Multinational Fin J [F]	1	2	0	2	0	1	1	1	.
Mustang J of Bus & Ethics [B]	5	0	0	0	0	0	0	0	.
NA J of Eco & Fin [F]	1	8	2	6	10	6	0	0	0.67
NA J of Fin & Ban Res [F]	1	0	0	0	0	0	0	0	.
NA Rev of Eco & Fin [F]	0	0	0	0	0	0	0	0	.
PSL Quar Rev [F]	0	1	0	0	1	0	0	0	.
Pac Acc Rev [A]	6	0	2	1	0	0	0	0	.
Pac-Bas Fin J [F]	0	9	3	4	1	5	0	0	0.41
Pan-Pac J of Bus Res [B]	0	0	0	0	0	0	0	0	.
Pensions: An Int J [F]	2	1	1	0	0	0	0	0	0.10
Private Equity Analyst [F]	0	0	0	0	0	0	0	0	.
Pub Budg & Fin [F]	1	18	4	1	6	3	0	1	0.60
Pub Fin Rev [F]	0	20	7	3	0	0	0	0	0.38
Qualitative Res in Acc & Mgt [A]	1	0	0	0	0	0	0	0	0.25
Qualitative Res in Fin Mark [F]	4	0	0	0	0	0	0	0	.
Quant Fin [F]	0	4	6	81	3	4	10	0	0.78
Quant Fin Letters [F]	0	0	0	2	0	0	0	0	.
Quar J of Fin & Acc [A]	1	5	4	0	3	0	0	0	.
Quar Rev of Eco & Fin [F]	2	12	9	7	17	4	6	7	0.56
Real Estate Eco [F]	0	3	3	2	0	3	0	1	2.55
Res in Acc Reg [A]	1	2	3	0	0	0	0	0	0.28
Res in Healthcare Fin Mgt [F]	0	1	0	0	0	0	0	0	0.00
Res in Int Bus & Fin [B]	1	4	2	5	2	4	0	1	0.46
Restaurant Bus [B]	0	0	0	0	0	0	0	0	0.11
Rev Esp de Fin y Contab [F]	2	8	0	0	0	0	0	0	0.14
Rev Eur de Dir y Eco de la Emp [B]	4	1	2	0	0	0	0	0	0.11
Rev Gestao Organizacional [A]	9	0	0	0	0	0	0	0	.
Rev de Contab & Contr [A]	18	1	0	0	0	2	0	0	.
Rev of Acc & Fin [A]	2	5	10	0	0	2	0	0	0.13
Rev of Acc Stu [A]	1	4	11	0	0	0	0	0	2.25
Rev of Asset Pricing Stu [F]	0	0	0	0	0	0	0	0	.
Rev of Beh Fin [F]	0	0	0	0	0	1	0	0	.
Rev of Bus [B]	6	2	4	0	1	0	1	0	.
Rev of Bus & Eco Res [B]	0	0	0	0	0	0	0	0	.
Rev of Bus & Fin Stu [B]	1	0	0	0	0	0	0	0	.
Rev of Bus Inf Sys [B]	1	0	1	0	0	0	0	0	.
Rev of Corp Fin Stu [F]	0	0	0	0	0	0	0	0	.
Rev of Deriv Res [F]	0	0	0	11	0	0	0	0	0.56
Rev of Dev Fin [F]	0	3	1	0	0	1	0	0	0.22
Rev of Fin [F]	0	2	1	0	0	0	0	0	3.45
Rev of Fin Eco [F]	1	6	7	6	4	0	0	0	0.25
Rev of Fin Stu [F]	0	27	18	22	0	0	0	0	12.61
Rev of Market Integ [F]	1	2	0	0	0	0	0	0	.
Rev of Pac Bas Fin Mark&Pol [F]	8	2	3	3	2	1	0	3	0.22
Rev of Quant Fin & Acc [F]	4	11	29	9	11	0	0	4	0.42
Revue Fin, Controle, Strategie [F]	2	0	1	0	0	0	0	0	.
Risk [F]	0	0	11	7	0	0	0	0	.
Risk & Decision Analysis [F]	0	0	1	0	0	0	0	0	0.13
Risk Mgt [F]	7	1	0	2	0	0	2	0	0.38
Risk Mgt & Ins Rev [F]	2	2	5	0	8	0	0	0	0.26
Romanian J of Eco For [F]	0	0	0	0	0	1	0	0	0.26
Ruffin Series in Bus Ethics [B]	0	0	0	0	0	0	0	0	.
Russian Fin Control Monitor [F]	0	0	0	0	0	0	0	0	.
SIAM J on Fin Math [F]	0	0	0	11	0	0	0	0	2.46
Scandinavian Actuarial J [F]	1	0	3	11	0	0	1	0	0.93
Scandinavian Int Bus Rev [B]	0	0	0	0	0	0	0	0	.
Seoul J of Bus [B]	3	3	1	0	0	0	0	0	.
Service Bus [B]	25	1	9	0	0	0	0	0	0.32
Small Bus Eco [B]	26	88	11	2	1	1	1	9	1.44
Small Bus Insti J [B]	1	0	0	0	0	0	0	0	.

Continued on next page

Journal	SPSS	Stata	SAS	MATLAB	RATS	EViews	SPlus	Limdep	SJR
Soc & Bus Rev [B]	3	0	0	0	0	0	0	0	.
Soochow J of Eco & Bus [B]	0	0	0	0	0	0	0	0	.
South Afr J of Bus Mgt [B]	0	0	0	0	0	0	0	0	0.16
South Asi J of Glo Bus Res [B]	1	0	0	0	0	0	0	0	.
Southern Bus Rev [B]	4	0	2	0	0	0	0	0	.
Southern J of Bus & Ethics [B]	1	0	1	0	0	0	0	0	.
Southwest J of Bus & Eco [B]	0	0	0	0	0	0	0	0	.
Spanish Rev of Fin Eco [F]	0	2	1	2	0	0	0	0	0.11
Stanford J of Law, Bus & Fin [F]	0	0	0	0	0	0	0	0	.
Strategic Fin [F]	1	0	8	0	0	1	0	0	.
Stu in Eco & Fin [F]	3	5	0	1	2	1	0	1	0.22
Sustain Acc, Mgt & Pol J [A]	1	1	1	0	0	0	0	0	0.35
Teach Bus & Eco [B]	1	0	0	0	0	0	0	0	.
Teach Bus Ethics [B]	3	0	1	0	0	0	0	0	.
The J of Real Estate Res [F]	4	8	8	4	2	4	3	1	1.71
Thunderbird Int Bus Rev [B]	13	1	7	1	0	1	0	0	0.32
Total Quality Mgt & Bus Exc [B]	14	0	1	1	0	0	0	0	0.57
Treasury & Risk [F]	0	0	0	0	0	0	0	0	.
Venture Capital [F]	6	4	0	0	0	0	0	1	0.35
World Ban Abstracts [F]	0	0	1	0	1	0	0	0	.
Zagreb Int Rev of Eco & Bus [B]	8	6	1	2	0	3	0	0	.

Note: The data are collected and are available upon request. Integers indicate the number of citations in journal articles that refer to a particular software during 1999–2013, for whatever reason. The notable omission is the R-project software due to the commonality in the keyword. The following software contained less than 200 citations each across all journals over 1999–2013 and are excluded from the table: Minitab, TSP, GAUSS, GLIM, SHAZAM, NLOGIT, Octave, GenStat, and Systat. Broadly defined, [A] indicates accounting, [B] indicates business, and [F] indicates finance.

14 Stata and political science

Guy D. Whitten
Department of Political Science
Texas A&M University
College Station, TX

Cameron Wimpy
Researcher
Fors Marsh Group
Arlington, VA

14.1 Introduction

Stata is today one of the two most popular statistics programs among political scientists. The rise of Stata to this status roughly parallels the rise of the Society for Political Methodology, which was started in 1983. When Stata was first launched, political science researchers were just starting to use models more complex than ordinary least-squares regression, and the need for a section of the American Political Science Association dedicated to statistical tools was not particularly obvious.[1] Today, the Political Methodology section is the American Political Science Association's second largest organized section.

Part of the success of the Society for Political Methodology has stemmed from its members' abilities to communicate the advantages and disadvantages of more complicated statistical modeling techniques to practitioners in the field. Through these efforts, the statistical sophistication of political science research has greatly increased. Of course, this increase in sophistication has led to a demand for software to help estimate more complicated models and diagnostic procedures. When the society first started, SAS and SPSS were the two dominant statistics programs among political scientists. However, the failure of these programs to include many of the popular new model classes led many researchers to explore alternative software packages such as Stata.

---

1. John Jackson, the first President of the Society for Political Methodology, wrote a nice account of the society's early days (Jackson 2012) that includes a discussion of their struggle to get the needed 100 signatures to start an organized section.

One of the major growth areas in political methodology has been in models of pooled time-series data. Neal Beck and Jonathan Katz's 1995 article on this subject was the eighth most cited article published in the *American Political Science Review* between 1945 and 2005 (Sigelman 2006). In fact, it was the only methodology article to place in the top twenty (Sigelman 2006), and it was the only article published in the 1990s to make the top ten on this list. Although we do not have any hard evidence, we strongly believe that the inclusion of `xtpcse`, the technique recommended by Beck and Katz (1995) in their famous article, and other `xt` tools has been a major force behind Stata's increased popularity among political science researchers.

Political science is one of the most popular disciplines studied at the world's two largest summer schools for social science research methods, the Inter-university Consortium for Political and Social Research (ICPSR) summer school at the University of Michigan and the Summer School in Social Science Data Analysis at the University of Essex.[2] When Stata was founded, SAS and SPSS were the most widely used programs in these schools. Today, Stata and R compete for this status. In 2014, there were four courses at ICPSR and one at Essex devoted to the use of Stata in social and political science research, with many other courses using Stata as the primary software program.

To provide a more systematic picture of the impact and popularity of Stata in political science, we present a graph in figure 14.1 that shows the number of articles that refer to Stata over time in the top two quantitatively focused political science journals, the *American Journal of Political Science* and *Political Analysis*.[3] This is a conservative estimate of the usage of Stata in these journals because there are many articles with evidence of certain types of figures and statistical procedures likely done with Stata that failed to specifically mention the program.

---

2. See Franklin (2008) for a discussion of the role of these summer programs, ICPSR in particular, and some data on their popularity in Political Science.

3. These numbers come from a search of the JSTOR database for the word "Stata".

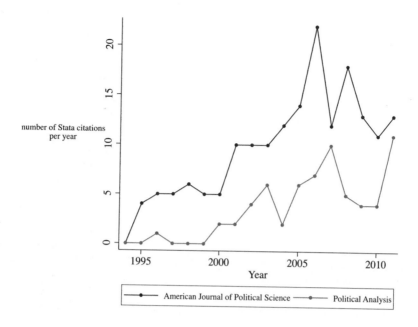

Figure 14.1. Stata citations on the rise

In recent years, there has been a trend of political science researchers using open-source software such as R or Python. These programs can be contrasted with traditionally closed programs such as SAS and SPSS. Stata provides a massive advantage in this area by offering, in a sense, the best of both worlds. On the one hand, open-source allows users to greatly expand and further develop the software. But, on the other hand, an absence of oversight means less quality control. Stata almost uniquely allows users to write software (in the form of ado-files) while maintaining full quality-control of the overall platform.

In some instances, user-written Stata programs have fundamentally changed the political science research landscape. Clarify, the postestimation software written by Gary King and colleagues (Tomz, Wittenberg, and King 2003; King, Tomz, and Wittenberg 2000), is one of the most prominent examples of this. Clarify allows users to readily use Monte Carlo simulations to produce "quantities of interest" that are easily interpreted and graphed. It was Stata's flexibility that allowed Clarify to have such a meaningful impact on research in political science and other disciplines.[4] The resulting software has allowed researchers to easily use simulation-based techniques to obtain more meaningful substantive inferences from their statistical models.

4. At the time of our writing, the Google Scholar citation counts were 2,061 for the article and 1,667 for the two different versions of the software documentation on Clarify.

Following King and his colleagues, many other political scientists have produced ado-programs, and articles on some of these programs have appeared in the *Stata Journal*. These include articles by Neumayer and Plümper (2010) on the setup of data for the estimation of spatial statistical models, Hicks and Tingley (2011) on the implementation of methods developed for mediation analysis, and Williams and Whitten (2011) on the use of dynamic simulations. The advent of the *Stata Journal* and the ever-growing user community on Statalist allow an unprecedented amount of communication among the developers and users of Stata. This level of community and support is unrivaled in the world of statistical software.

Stata also shines in another important area of political science research: the repro-duction of published results. The ability to reproduce research is an essential component of the scientific process.[5] By stressing the use of the code-documenting do-files (`.do`), Stata helps users create work that is reproducible even when that is not the goal. By including log files and well-documented do-files, researchers using Stata can easily reach the gold standard of reproduction. Stata goes a step further for users preferring a "point and click" style graphical user interface. When a user clicks menu-driven commands in Stata, the code needed to reproduce that command is printed in the results window. These commands can be recorded via a log file or copied elsewhere for future reference and reproduction. Stata's version-control system can further ensure that researchers can reproduce results as they were originally created by their particular version of Stata. Few programs can match Stata's flexibility when it comes to easily documenting code and reproducing results.

The flexibility and adaptability of Stata bodes well for its future as an integral part of political science research. Recent programs from the annual meetings of the Society for Political Methodology are usually a good predictor of where the discipline is heading in terms of its use of statistical methodology. If we look at these programs, we can see articles on a broad range of topics including the use of Bayesian methods of inference, experimental and quasi-experimental research designs, and spatial statistical models. Each of these topics involves techniques that can be easily implemented in Stata.

14.2 Acknowledgment

We thank Thiago Silva for collecting the JSTOR data on references to Stata in political science journals.

References

Beck, N., and J. N. Katz. 1995. What to do (and not to do) with time-series cross-section data. *American Political Science Review* 89: 634–647.

---

5. The entire January 2014 issue of *PS: Political Science & Politics* was devoted to reproduction and transparency in political science.

Franklin, C. H. 2008. Quantitative methodology. In *The Oxford Handbook of Political Methodology*, ed. J. M. Box-Steffensmeier, H. E. Brady, and D. Collier, 796–813. Oxford: Oxford University Press.

Hicks, R., and D. Tingley. 2011. Causal mediation analysis. *Stata Journal* 11: 605–619.

Jackson, J. E. 2012. On the origins of the Society. *Political Methodologist* 19: 2–7.

King, G., M. Tomz, and J. Wittenberg. 2000. Making the most of statistical analyses: Improving interpretation and presentation. *American Journal of Political Science* 44: 341–355.

Neumayer, E., and T. Plümper. 2010. Making spatial analysis operational: Commands for generating spatial-effect variables in monadic and dyadic data. *Stata Journal* 10: 585–605.

Sigelman, L. 2006. The American Political Science Review Citation Classics. *American Political Science Review* 100: 667–669.

Tomz, M., J. Wittenberg, and G. King. 2003. Clarify: Software for interpreting and presenting statistical results. *Journal of Statistical Software* 8: 1–30.

Williams, L. K., and G. D. Whitten. 2011. Dynamic simulations of autoregressive relationships. *Stata Journal* 11: 577–588.

14.3   About the authors

Guy D. Whitten is a professor in the Department of Political Science at Texas A&M University. He is also the director of the European Union Center and the director of the Program in Scientific Political Methodology.

Dr. Whitten's primary research and teaching interests are political economy, public policy, political methodology, and comparative politics. Much of his published research has involved cross-national comparative studies of the influence of economics on government popularity and elections. His recent research in public policy includes work on the politics of defense spending in parliamentary democracies. Dr. Whitten has also published various influential works on using statistics to make inferences in political science. Together with Paul Kellstedt, Dr. Whitten has written a textbook titled *The Fundamentals of Political Science Research* (Cambridge University Press 2009, 2013). He was a coeditor of *The Political Methodologist* and currently serves on the editorial boards of *The American Journal of Political Science*, *Electoral Studies*, and *Political Science and Research Methods*.

Dr. Whitten is a frequent instructor of short courses on statistical methods for social scientists and other analysts of observational data at many international universities and research centers. These include the Essex School in Social Science Data Analysis at the University of Essex (United Kingdom), the International Political Science Association's

Summer Schools on Concepts, Methods and Techniques in Political Science (in both Sao Paulo and Singapore), Concordia University (Montreal), and the International Crane Foundation (Cuba). His methodological specialties include models of discrete dependent variables, multilevel data, and dynamic processes.

Cameron Wimpy is a researcher at Fors Marsh Group in Arlington, VA. He works in science and analysis across a variety of research teams and provides consultation and training on Stata and related statistical procedures. A political scientist, his substantive research focuses are comparative political economy and quantitative research methods. His publications have appeared in journals such as *Political Science Research and Methods* and *Electoral Studies*. He has conducted survey fieldwork in several African and Middle Eastern countries and has presented at professional conferences around the world. He received his PhD from Texas A&M University in College Station, TX.

15 New tools for psychological researchers: Psychological statistics and Stata

Stephen Soldz
Boston Graduate School of Psychoanalysis
Brookline, MA

15.1 Introduction

Those familiar with quantitative data analysis are well aware that fields differ in the kinds of data they produce, the types of questions they attempt to answer, and the statistical techniques they traditionally use to address those questions. Psychologists and psychological problems played major roles in the early development of statistics and the development of many techniques—for example, exploratory factor analysis, multidimensional scaling, and structural equation modeling. Yet, for most practical psychological researchers, data analysis involved the application of only a small collection of standard, and relatively simple, statistical procedures until fairly recently.

Not so long ago—okay, it was quite a while ago—most psychological questions in routine research involved the associations between variables or group differences often resulting from experiments. Practical psychological statistics consisted largely of t tests and analysis of variance (ANOVA) for experiments and other quantitative group difference questions and correlations and regression for association and prediction, with the odd chi-squared thrown in for the occasional contingency table.

Other more advanced researchers learned "multivariate statistics", which consisted of exploratory factor analysis for item analysis and test construction along with a hodge-podge of techniques that expanded upon correlations for associations, t tests, and ANOVA and allowed exploration of group differences when these differences were expressed by multiple variables. Canonical correlation, though little used, was taught as a generalization of the bivariate Pearson correlation. Discriminant function analysis and multivariate analysis of variance were the central techniques. Cluster analysis, another way of exploring group differences, was often used, as sometimes was nonmetric multidimensional scaling, although uses of the latter in practical research were rare.

While psychology always had many very sophisticated advanced quantitative researchers who published in journals such as *Psychometrika*, *Psychological Bulletin*, and *Psychological Methods*, their work exerted little influence on practical research. In general, the mathematical level of most researchers was quite low. In my doctoral program in the 1980s, for example, none of the faculty, including those teaching statistics and research methods, had any background in basic calculus or linear algebra. While the multivariate course (which was taught in the statistics department in my school but was rarely taken by psychologists) mentioned matrices, students were reassured that they would not need to actually understand them. Eigenvalues were numbers on the factor analysis printout and were supposed to be greater than 1.0 if they mattered. Even the extremely influential psychological methodologist Jacob Cohen (1992, 409) had almost no mathematical knowledge and viewed this as an advantage:

> With little more statistics than a graduate year devoted largely to analysis of variance and chi-square and no more math at my command than some half-remembered high school algebra, proofs requiring the differential calculus and matrix algebra were quite beyond me. My method was to get a hunch, try it out with three or four data sets, and accept it if it worked on all the sets and reject it otherwise [...].

> I had no choice but to write as I thought and as I thought most psychologists thought. [...] I believe that at least part of the reason for the regression articles many citations is exactly the nonmathematical, verbal-intuitive (fuzzy), redundant style to which the sophisticated methodologists objected. My desire to communicate in the proper technical language is a defect that has its virtues, because it is widely shared among psychologists and their ilk. The mathematically facile psychologists enormous advantage in formulating novel methods and communicating with each other is lost when they address an audience of behavioral scientists, a very bright group of people not known, however, for their competence in mathematical statistics.

Alas, the divide between practical psychology researchers and the mathematically sophisticated has been only partially removed, though it is more common for major funded research studies to have a sophisticated consultant who can speak that arcane language that is still so foreign to most researchers.

When I was in grad school, statistical analysis was still largely conducted on mainframes, primarily using SPSS, though a few of us taught ourselves SAS. But even familiarity with these programs did not mean facility with conducting more complex analyses. Much, if not most, analyses were conducted by graduate students or research assistants who were often largely self-taught. Only the most well-funded research projects had a person advanced in data analysis. When word got around among the faculty that I had figured out how to get SPSS to conduct repeated-measures multivariate analysis of variance, I was drafted to teach the arcane syntax to other grad students conducting statistical analyses, as they had been stumped by this.

Practical psychological statistics started to change in the mid-1970s with Cohen's (1968) explanation (elaborated in his classic textbook [Cohen et al. 2003]) that many of the diverse techniques of psychological statistics were actually special cases of the general linear model and that statistical models look a lot like the multiple regression that we psychologists were becoming familiar with.

If you should say to a mathematical statistician that you have discovered that linear multiple regression analysis and the analysis of variance (and covariance) are identical systems, he would mutter something like, "Of course—general linear model", and you might have trouble maintaining his attention. If you should say this to a typical psychologist, you would be met with incredulity or worse.

It is readily understandable that psychologists would find strange the claimed equivalence of multiple regression and the fixed-model analysis of variance and covariance. The textbooks in "psychological" statistics treat these matters quite separately, with wholly different algorithms, nomenclature, output, and examples (Cohen 1968, 426). This news was received by those interested in practical research methodology as a revelation and gradually penetrated the thinking of many psychological data analysts over the next 15 years or so. However, the transition from data analysis as being the application of disparate techniques to applications of general models is still in process.

In retrospect, the concept of models was central for introducing a wider range of statistical approaches to psychological researchers and in opening up more research techniques that were previously associated with other disciplines. Previously, to determine whether a psychological treatment affected the time till a recovered depressed individual relapsed to depression, we might categorize the relapse times and conduct a chi-squared. In the model era, we developed access to survival analysis, previously a property of biostatisticians, or its sociological twin, life history analysis.

Coinciding with the gradual transition to a model-based framework for thinking about statistical analysis, mainstream psychological research moved to exploring more questions than variable associations, group differences, or measurement development. Psychological methodologist Wayne Velicer (2014) describes new types of questions psychologists are addressing concerning the identification of patterns and mechanisms of change over time in individuals and groups, which leads to a consideration of dynamic variables and mediators. Velicer, like many other researchers in psychology and elsewhere, calls attention to problems with the traditional null hypothesis significance testing and supports the emerging focus on effect-size estimation, power analysis, and meta-analysis.

The two types of software traditionally used by psychologists—SPSS and SAS—are still based in the era well described in the Cohen quote above (1968, 426), an era when statistical analysis was a collection of independent techniques. While both packages include a multitude of "modern" techniques, such as multilevel modeling or generalized estimating equations, the logic of the packages encourages users to think of them as discrete techniques. This lack of a conceptual framework was of little importance during the era in which psychologist researchers overwhelmingly used few classical procedures.

Stata, in contrast, was originally developed with the concept of statistical models at its core. Thus, the syntax is consistent across types of models, as is the output. Most models use regression-like syntax and regression-like output tables. The general linear model becomes generalized to generalized linear models, with largely analogous structure. These, in turn, become generalized to include both multilevel features and structural equations as in the user-written Stata program generalized linear latent and mixed models (Rabe-Hesketh, Skrondal, and Pickles 2004; Skrondal and Rabe-Hesketh 2004). Other models, including many specialized models in base Stata, also follow similar syntax and generate analogous output.

There is such consistency across classes of models that users have been able to develop "wrappers" to turn the output from a range of quite varied models into exportable tables. This consistency across models is one of several Stata features that makes the program especially easy for researchers to learn and use and aids the understanding of new types of analytic approaches. However, for those used to output in specialized formats, such as ANOVA tables, these are available as well, as are procedures that do not fit within the model framework, such as cluster analysis.

Psychologists have also started to recognize that much of their data do not meet the independence of observations assumptions of the classical statistical techniques and that many datasets are, in fact, multilevel with primary units—for example, students nested in higher-order units such as classrooms and schools. Further, psychological data are often longitudinal in character, which also builds in correlations between observations (Soldz 2006). This recognition has led to increased interest in using analytic techniques that properly account for this nonindependence or nesting, such as multilevel modeling and generalized estimating equations (Fitzmaurice, Laird, and Ware 2011; Hardin and Hilbe 2013; Singer and Willett 2003). Stata was one of the first major packages to include various techniques for longitudinal and other nonindependent data, and it has continually expanded the options for these complex yet vital types of analysis.

15.2 Choosing Stata

As a researcher, my attraction to Stata goes well beyond the ease and consistency of its model-based framework, however. For me, the most important Stata feature is its data-management capabilities. In my experience, tasks that take hours or even days to accomplish in SAS or SPSS often take only minutes in Stata. Especially powerful is Stata's macro substitution system, which, combined with its built-in ability to loop over variables or any other objects, allows the user to easily accomplish tasks that are likely possible yet difficult in SAS or SPSS. This difference between the packages was driven home to me when I was faced with a complex renaming task for over 500 variables from a longitudinal study. A SAS user at the time, I spent several days pouring over the SAS Macro and other manuals, figuring that there had to be a way to do it using SAS. While I am sure there is such a way, I never found it. I was starting to teach myself Stata at the time, so I turned from SAS to Stata. Fifteen minutes later, the task was accomplished. After I left my job position at the time, which required SAS, I never used SAS again. Stata was just so much easier to get to do what I wanted.

At that point (2002), Stata maintained its model approach and was weak in the multivariate techniques traditional in advanced psychological research, as well as in multivariate exploratory data analysis procedures for exploring data structure, such as multidimensional scaling or correspondence analysis. When I presented at the Stata Users Group meeting in 2003 and highlighted these weaknesses (Soldz 2003), the president of StataCorp, Bill Gould, immediately came over and asked for my help in removing these weaknesses. In the following year or so, Stata produced an excellent suite of multivariate analysis procedures that removed these obstacles, hence, encouraging adoption of Stata by psychologists. Simultaneously, Stata worked at improving its package's capability to conduct other types of analyses common among contemporary psychologists, such as complex experimental designs, structural equation modeling, and multilevel models. Stata's capabilities in these areas are impressive and, in many instances, rival that of standalone programs that cost as much as or more than the complete Stata package.

Bill Gould's response to me illustrates another major strength of Stata: its support for and responsiveness to users. Many years ago, when my research group had trouble with SPSS' then data-entry module, we called customer service several times and were stonewalled, only to have them admit after six weeks that we had lost over a month's work doing something—reordering variables—that the documentation nowhere said we could not do. In contrast, StataCorp staff regularly admit bugs on what used to be the Statalist listserver (now transformed into a forum, www.statalist.org/), and they often discuss potential solutions with complainers and issue updates every few weeks while also documenting all bugs, their fixes, and what is not fixed. The Statalist forum also provides extensive support from highly experienced users that nicely supplements support provided directly by StataCorp.

A final strength of Stata for advanced researchers is the ease of writing user programs to conduct types of analyses or other procedures (for example, formatting output in novel ways or interfacing with external programs) not included in official Stata. While most major statistics packages have this capacity in some form, most maintain a clear separation between in-built and user-written procedures. The latter are written in a specialized programming or macro language far different from the native language of the package. This is not the case in Stata. While certain core capability and legacy programs are written in C, most procedures are written in the same programming language available to users and are treated the same whether they come with Stata or are user written. The process of writing add-on programs is simplified because Stata saves an enormous number of its statistical results, making them accessible to users. This process has expanded in the past decade as Stata added a full matrix programming language, Mata, which includes the LAPACK set of linear algebra algorithms (LAPACK—Linear Algebra PACKage 2013), and began a process of rewriting many of their C procedures in Mata. All the procedures written in Stata's traditional language or Mata are readable by users, allowing access to internals and their use as models for user-written code. There is an extensive library of user-written commands for all kinds of procedures that is directly accessible from within Stata, and these commands can be found by Stata's search capability. Thus Stata is second only to R in its openness to users; in a sense, it is a hybrid system combining features of traditional software and open-source software.

My experience with statistical software has covered many of the options. I started using SPSS in graduate school. I then used SAS when working on my postdoc at Harvard Medical School; at that time, SAS was still accessed remotely via a Unix server and error logs were output in a computer center miles away, so I was forced to debug without any information on my errors. Thus, when I got my first PC, I switched back to SPSS only to abandon it again after the unfortunate episode described above when SAS became available on PC. I used SAS primarily for over a decade, though I experimented with various alternatives—SYSTAT, Statistica, and S+—however, none of them had the depth of statistical power in SAS. R, while having that depth, was hard to learn and it was difficult to retain the knowledge to use it productively during periods when I did not need it for several months, as is typical of my episodic immersion in data analysis. I have never had problems picking up Stata again after a hiatus.

I was introduced to Stata in 1999 by a local Boston expert, Michael Blasnik, who used it on enormous datasets of hundreds of thousands of utility customers. As Director of Research at a nonprofit research organization, I hired Michael to help with analysis of a substance abuse survey we conducted for the Massachusetts Department of Public Health. Michael impressed with his ability to use Stata to automate the production of dozens upon dozens of complex tables using a program he wrote for the purpose (Blasnik 2001), allowing him to do in forty hours what would have taken our staff many weeks of effort.

Over the next couple of years, I experimented with Stata while using SAS as my primary tool. At the time, I was conducting data analysis of surveys with classrooms as the primary sampling units, which traditionally required specialized software. At that point, both Stata and SAS had the capability to analyze surveys. However, I found Stata's procedures far easier to learn and use, as I did Stata's advanced data management facilities. The switch to Stata was easier when I no longer had access to an institutional site license for SAS and found that the cost of a permanent license for Stata was comparable to one year of a license for SAS. The more I used Stata, the more enamored I became of its power and ease of use.

As a psychologist, I have felt rather isolated in my use of Stata. Most of my psychologist researcher colleagues use SPSS, which they learned in school, with some using SAS and adventurous members of the younger generation using R. While all of these packages have excellent capabilities and different strengths, based on my experience, I think that many of my colleagues might be pleasantly surprised by what Stata has to offer if only they would give it a try.

References

Blasnik, M. 2001. svytabs: Stata module to create tables for export combining multiple survey tabulations. Statistical Software Components S418301, Department of Economics, Boston College. http://ideas.repec.org/c/boc/bocode/s418301.html.

Cohen, J. 1968. Multiple regression as a general data-analytic system. *Psychological Bulletin* 70: 426–443.

———. 1992. Fuzzy methodology. *Psychological Bulletin* 112: 409–410.

Cohen, J., P. Cohen, S. G. West, and L. S. Aiken. 2003. *Applied Multiple Regression/Correlation Analysis for the Behavioral Sciences*. 3rd ed. Mahwah, NJ: Lawrence Erlbaum.

Fitzmaurice, G. M., N. M. Laird, and J. H. Ware. 2011. *Applied Longitudinal Analysis*. 2nd ed. Hoboken, NJ: Wiley.

Hardin, J. W., and J. M. Hilbe. 2013. *Generalized Estimating Equations*. 2nd ed. Boca Raton, FL: CRC Press.

PACKage, LAPACK.-L. A. 2013. LAPACK—Linear Algebra PACKage. http://www.netlib.org/lapack/.

Rabe-Hesketh, S., A. Skrondal, and A. Pickles. 2004. GLLAMM manual. Working Paper 160, Division of Biostatistics, University of California–Berkeley. http://www.bepress.com/ucbbiostat/paper160/.

Singer, J. D., and J. B. Willett. 2003. *Applied Longitudinal Data Analysis: Modeling Change and Event Occurrence*. Oxford University Press: Oxford.

Skrondal, A., and S. Rabe-Hesketh. 2004. *Generalized Latent Variable Modeling: Multilevel, Longitudinal, and Structural Equation Models*. Boca Raton, FL: Chapman & Hall/CRC.

Soldz, S. 2003. Multivariate data exploration with stata: Evaluation and wish list. Boston, MA: 2nd North American Stata Users Group meetings. http://www.stata.com/meeting/2nasug/soldz.ppt.

———. 2006. Models and meanings: Therapist effects and the stories we tell. *Psychotherapy Research* 16: 173–177.

Velicer, W. F. 2014. New directions in behavioral statistics. http://www.apa.org/divisions/div5/pdf/April14Score.pdf.

15.3   About the author

Stephen Soldz is a psychologist, psychoanalyst, and public health researcher in Boston. He is the director of the Center for Research, Evaluation, and Program Development at the Boston Graduate School of Psychoanalysis. He was an adjunct assistant professor of psychology (psychiatry) at Harvard Medical School, and he has taught at the University of Massachusetts Boston, Boston College, and Boston University.

Dr. Soldz is an expert on research methodology and has conducted numerous research and evaluation studies and published dozens of papers on such topics as psychotherapy process and outcomes, personality development and pathology, research methodology, substance abuse, and tobacco control. He has consulted to local, state, and federal government organizations and managed-care organizations on research and evaluation issues. Dr. Soldz coedited *Reconciling Empirical Knowledge and Clinical Experience: The Art and Science of Psychotherapy*, published by American Psychological Association Books in 2000.

16 A short history of Stata on its 30th anniversary

Nicholas J. Cox
Durham University
Durham, UK

16.1 Introduction

In 2005, the *Stata Journal* included a special issue on the 20th anniversary of Stata, including a capsule history of Stata by the present author (Cox 2005). It was based mostly on talks given by Bill Gould at Stata Users Group meetings in London and Boston in 2004 and on conversations with him, other developers of Stata, and various users over the previous decade.

Attempting an update runs the usual risks of recent history, and some more. Writing from outside the company, and even the country, that produces Stata can be puffed as providing a more impartial perspective, but that double distance makes many important ideas and facts difficult, if not impossible, to discern. The growth of Stata is an even more obvious hazard to an amateur (and very part-time) historian. Even being aware of, let alone documenting, the major changes in Stata and the Stata community is a tough challenge. But enough excuses: much of the point of the present collection is that each author has a useful and complementary perspective on things and people Stata.

Table 16.1 gives a compact list of Stata's release dates, the main stations on its journey over 30 years, with more to come. Careful readers will note that some dates are corrected from those given in Cox (2005).

Table 16.1. Releases of Stata

1	January 1985	8	January 2003
1.1	February 1985	8.1	July 2003
1.2	May 1985	8.2	October 2003
1.3	August 1985	9	April 2005
1.4	May 1986	9.1	September 2005
1.5	February 1987	9.2	April 2006
2	June 1988	10	June 2007
2.05	April 1989	10.1	August 2008
2.1	August 1990	11	July 2009
3	March 1992	11.1	June 2010
3.1	August 1993	11.2	March 2011
4	January 1995	12	July 2011
5	September 1996	12.1	January 2012
6	January 1999	13	June 2013
7	December 2000	13.1	October 2013

16.2 Californian creation

Stata was started in California in the mid-1980s. In 1984, a UCLA graduate named William Gould was part of a small but successful business selling computer time on mainframes called Computing Resource Center and based in Santa Monica. At the time, personal computers (PCs) that could be taken seriously were being launched on the market, and it was clear that they were no longer just toys for hobbyists. Further, several people were writing statistical programs directly intended for PCs, such as Systat or MicroTSP. Stata was based on Bill's bold assumption, the notion that (with help) he could write a program that could stand comparison with any on the market.

Stata started life within the company as DiAL, a name taken from a never-released project; a logo existed and several spare folders were lying around the office. At one time, the name S was considered, until Bill Rogers pointed out that a group at Bell Labs had got there first. The name was changed to Stata just before release. In some early documentation, it was shouted out in all capitals as "STATA", but the presently used form emerged quickly.

Bill Gould and Sean Becketti spent a year writing the first version of Stata. It was written, then as now, in the C programming language. Anyone who knows C will recognize that the syntax of Stata is influenced by the syntax of C, although many details (including preferences for lowercase names and terseness of output) were also generic to the Unix operating system. In that first year, much time was spent not writing code, but arguing about the design and trying very hard to get it right, thus establishing a tradition of discussing even simple things like command names, until, ideally, a solution emerges by consensus among the developers.

Stata 1.0 was a small program that could not claim to cover all of mainstream statistics any more than its competitors did. It could more fairly be described as a regression package with data-management features. Stata 1.0 existed only as a program for PCs running the DOS operating system. Most of its 44 commands (table 16.2) will look familiar to present users. From time to time, there have been facetious suggestions that Stata 1.0 should be reissued (apart from its bugs, naturally). At least the manual could be lifted without difficulty.

Table 16.2. Stata 1.0 and Stata 1.1

append	dir	infile	plot	spool
beep	do	input	query	summarize
by	drop	label	regress	tabulate
capture	erase	list	rename	test
confirm	exit	macro	replace	type
convert	expand	merge	run	use
correlate	format	modify	save	
count	generate	more	set	
describe	help	outfile	sort	

Stata 1.0 had `int`, `long`, `float`, and `double` variables; it did not have `byte` or `str#`.

Stata 1.1 was a release to fix bugs, but 1.2 brought a new menu system and better online help. Then Stata 1.3 brought real graphics on IBM CGA and Hercules Monochrome cards through an add-on called Stata/Graphics. In that version, a `program` command was implemented, although its existence was not revealed until later. Making Stata extensible so that users could add their own commands and the software could begin growing from a package into a language was undoubtedly one of the most important steps in its history.

For the first few years of Stata's history, there were releases of various kinds, both new versions and extra sets of Stata programs (sometimes sold as "kits", or later as support disks), every few months. Among the more notable additions were `anova`, `logit`, and `probit` in Stata 1.5. Some users had proved oddly resistant to an initial bias that ANOVA was just a roundabout way of doing regression.

A move beyond DOS to an extra operating system, Sun/Unix, came in 1988. 386/ix and HP/9000, DEC RISC, IBM RS/6000, DEC Alpha, and other variants of Unix (later of Linux) followed. The first Stata for Macintosh (written by Bill Rogers) was released in 1992. Since then, other versions for the Macintosh have been developed, overseen by Chinh Nguyen, most recently for OS X. Maximizing platform independence has long been a major principle for Stata.

Stata 2.0 was a major new release, with revised graphics now integrated into the Stata executable and not implemented through add-ons, and with string variables, survival analysis (Cox and Kaplan–Meier), and stepwise regression. Stata 2.05 saw the documentation rebound in a new format. Stage 1.0 was released shortly afterwards. This stage was an interactive graphics editor that was written in one month and never updated; it was ported to Linux much later but in other respects, it was not modified. (Users had to wait until Stata 10 for a new interactive graphics editor that matched the new graphics introduced in Stata 8.)

The first Stata text was *Statistics with Stata*, by Lawrence Hamilton (1990) (see also Hamilton [2005]). This was another sign of maturity, and it fed awareness and sales of Stata to mutual benefit. Teaching of statistics with Stata, or of Stata itself, has never been the first concern of Stata's developers, who see Stata primarily as software for research, but there has always been strong and positive support for people who write texts linked to Stata. Among many subsequent texts and teaching ventures, an early venture that must be mentioned is StataQuest, a joint project with Duxbury Press. The accompanying text was written by Ted Anagnoson and Richard DeLeon; see also Anagnoson (2005).

Later copies of Stata were sold through distributors in various countries outside the United States. Peter Hedström of Metrika Consulting in Sweden was the first Stata distributor, and the late Ana Timberlake of Timberlake Consultants in Britain was the second. For an appreciation of Ana, see Timberlake and Cox (2010). The number now is much larger, as can be seen from the StataCorp website.

In April 1990, Computing Resource Center exited the mainframe service bureau business. The sideline product had become the mainline product.

Stata 2.1 saw other major additions: byte variables, factor analysis, and ado-files. `reshape`, previously distributed in a kit, also become integrated into Stata. The support for ado-files has been the most important of these additions over the long term. That is, a program could be not only defined by a file but also automatically loaded into memory when the user needed it. Its addition at this time was in a way an accident, as Bill Gould has explained (Newton 2005).

Another venture at this time was the introduction of the *Stata Technical Bulletin* (STB) as a publication issued 6 times per year, initially under the editorship of Joe Hilbe (see Hilbe [2005]). This continued publication (later under the editorships of Sean Becketti and Joe Newton) for just over 10 years (61 issues) and, in a strong sense, has continued in a new guise as the *Stata Journal* since late 2001. Table 16.3 gives a list of people who have served as members of the editorial boards of the two journals.

Table 16.3. Editors and Associate Editors, STB and *Stata Journal* (Associate Editors unless otherwise stated)

Joseph Hilbe	STB Editor 1991–1993, 1993–1994
J. Theodore Anagnoson	STB 1991–1994
Richard DeLeon	STB 1991–1994
Paul Geiger	STB 1991–1994
Richard Goldstein	STB 1991
Stewart West	STB 1991–1994
Lawrence C. Hamilton	STB 1991–1994
Sean Becketti	STB Editor 1993–1996
Francis X. Diebold	STB 1994–2000
Joanne M. Garrett	STB 1994–2001, SJ 2001–2007
Marcello Pagano	STB 1994–2001, SJ 2001–
James L. Powell	STB 1994–1998
J. Patrick Royston	STB 1994–2001, SJ 2001–
H. Joseph Newton	STB Editor 1996–2001, SJ Editor 2001–
Nicholas J. Cox	STB 1998–2001, SJ Executive Editor 2001–2004, SJ Editor 2005–
Jeroen Weesie	STB 2000–2001, SJ 2001–
Christopher F. Baum	STB 2001, SJ 2001–
Rino Bellocco	SJ 2001–
David Clayton	SJ 2001–2007
Charles Franklin	SJ 2001–2007
James Hardin	SJ 2001–
Stephen Jenkins	SJ 2001–
Jens Lauritsen	SJ 2001–
Stanley Lemeshow	SJ 2001–
J. Scott Long	SJ 2001–
Thomas Lumley	SJ 2001–2009
Sophia Rabe-Hesketh	SJ 2001–
Philip Ryan	SJ 2001–
Allan Gregory	SJ 2002–
Jeffrey Wooldridge	SJ 2002–
Mario A. Cleves	SJ 2003–
Roger Newson	SJ 2003–
Mark E. Schaffer	SJ 2003–
William D. Dupont	SJ 2004–
Ulrich Kohler	SJ 2004–
Nicholas J.G. Winter	SJ 2004–
Ben Jann	SJ 2005–
A. Colin Cameron	SJ 2006–
Nathaniel Beck	SJ 2008–
Maarten L. Buis	SJ 2008–
David Epstein	SJ 2008–
Frauke Kreuter	SJ 2008–
Austin Nichols	SJ 2008–
Peter A. Lachenbruch	SJ 2010–
Philip Ender	SJ 2012–
Ian White	SJ 2013–

The STB, as printed journal plus diskette, served as a medium for publication of extra-official programs between releases, superseding the irregularly issued kits and support disks. Increasingly, and more importantly, it served as a medium for users to make their own programs public and thus to formalize the existence of a genuine user community. Researchers could not only benefit from others' programming labor, but also gain some credit from the publication of their own work. It is striking how many of the additions to "official Stata" (that which the company sells and supports) in the 1990s had roots, directly or indirectly, in user-written programs. `glm`, to which both Joe Hilbe and Patrick Royston (see Royston [2005]) made outstanding contributions, is just one example among several.

16.3   Texan transpose

In August 1993, the company moved to College Station, Texas, and became Stata Corporation. More recently, its name has been tweaked for legal reasons to StataCorp LP. In 2000, the company moved out of what had become very cramped quarters near Texas A&M University to a new building on its own campus. In 2005, a second building was completed.

For a few years from Stata 3.0 on, the main development thrust was to extend the corpus of regression-like modeling commands. Among many others were added flavors of regression for limited, categorical, bounded, and multiple dependent variables, robust and quantile regression, and support for non-normal distributions. Central to this set of commands was an engine for maximum-likelihood-based modeling that emerged as `ml` and thus permitted users to program their own modeling commands. Later came much more for panel data, time series, survey estimation, and cluster analysis (and much more yet).

A development with which the company had little to do, at least initially, was the Statalist listserver, started by David Wormuth in August 1994. It was based on hardware and software running at Harvard University. After a few years, Marcello Pagano took over moderation of the list with assistance from Nicholas Cox, who maintains the FAQ. Statalist quickly established its own traditions. It has been a most valuable forum for discussions about Stata, and statistical science generally, and has encouraged the writing and dissemination of many hundreds of Stata programs. Some were transient— but likely the solution to somebody's problem—but many have proved of widespread and lasting value.

Many emails on the Statalist included listings of programs, which allowed interested users to copy and paste programs into their own files. Some of the longer programs had to be split into multiple postings! Note that Statalist did not support attachments until its relaunch as a web forum in March 2014. Installing programs in this way was bound to be onerous and error-prone. A key shift occurred in 1997 when Christopher (Kit) Baum volunteered to put Stata programs sent to him by users within the Statistical Software Components (SSC) archive he was maintaining. This central searchable archive of user-written programs has proved very popular. Although some users continue to make their programs available through personal websites, the majority of user-written Stata programs in the public domain (and not published through the STB or the *Stata Journal*) are found within the SSC archive. This was a development with which the company had initially no involvement. Later, however, they added an `ssc` command to Stata based on programs by Baum and Cox, thus giving users a handle to install programs directly without the intermediary of a browser.

Stata 4.0 saw the first port to Microsoft Windows (3.1 to be precise). At the time, PC users were moving in large numbers from DOS to Windows, and there was a strong case that Stata should follow suit. However, it was difficult for the developers to see exactly how a Windows interface could be consistent with the command language that was so utterly central to Stata's design.

Various possibilities were entertained in an internal project, but the company would not market until they were happy with a new design. One day, Bill Sribney, then working as a developer, came into work with a vision of Command, Results, Review, and Variables windows. This immediately appealed, and it remains at the center of all Stata's windowed interfaces, even for Macintosh and Unix. After Stata's move to Windows (3.1) followed Stata for other flavors of Windows (95, NT, XP, and so on); here Alan Riley has overseen development.

Stata Users Group meetings began in 1995, with the first held in London and organized by Ana Timberlake. Figure 16.1 gives a graphical overview of the meetings from 1995 to the end of 2014. For those interested in cities rather than countries, the list of venues in order of frequency starts with London, followed by Boston, Berlin, Mexico City, Stockholm, and Madrid. The major meeting each year in the United States is now known as the Stata Conference.

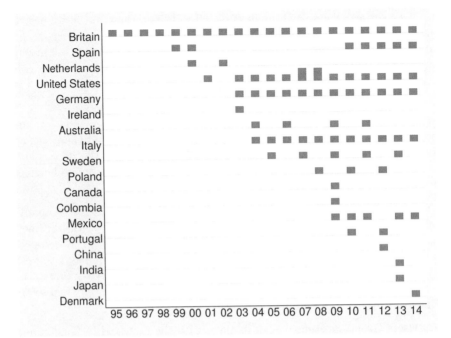

Figure 16.1. Stata Users Group meetings in various countries from 1995 to December 2014. Each bar represents either one meeting in each country in each year or, in two instances, two meetings in the United States in the same year.

The very first meeting in London established a pattern followed by all subsequent meetings. The logistics of times, places, registration, food and drink, and so forth are always handled by the distributor (in the United States, by StataCorp). The program is always handled by users. User presentations are typically combined with a "Report to users" from a senior Stata developer. The formal meeting always ends with a "Wishes and grumbles" session (this was Ana Timberlake's expression, which has since been copied across the world) at which users shout out requests and complaints and are met with "Yes", "No", or "Maybe". In some recent meetings, there have been longer review talks and training sessions, which have proved popular.

It is difficult to encapsulate what Stata Users Group meetings mean for those concerned. Evidently, Stata's work is highly computer based, so most interaction between (and among) users, developers, and technical support can be through email and the Internet more generally. Equally evidently, there can be no real substitute for face-to-face interaction, and many important developments have grown from these meetings. Sophia Rabe-Hesketh spoke about gllamm at a Users Group meeting, thus launching one of the most versatile and widely used user-written programs. Users too report on exchanges, collaborations, and friendships that would not have arisen from disciplinary conferences, and they often remark that the meetings are more pleasant, and even more useful, than such conferences. Which disciplines? I could not possibly comment.

By industry standards, Stata was a bit slow to respond to the Internet and all the possibilities it created, but the software can fairly claim to have caught up. The company's first net-based venture was into what were called NetCourses. James Hardin, then a developer, now at the University of South Carolina, was most active in supporting the NetCourse idea, despite a widespread view in the company that it would not work. Some NetCourses have led to books released by Stata Press, StataCorp's publishing arm, the first being a text on maximum likelihood by Bill Gould and Bill Sribney (1999), most recently extended to its fourth edition by Gould, Jeff Pitblado, and Brian Poi (2010).

The website *www.stata.com* was launched in 1996 and has since become central to the Stata enterprise. It was somewhat later, in Stata 6, that Stata itself became web-aware, allowing reading of datasets (and other files) across the net and `updates` of Stata without the need for distributing diskettes or CDs.

The Internet also had long-term consequences for the STB. When it started, distribution of diskettes by mail was the most practical way for most users to acquire copies of new programs. But the rise of email and of websites made that role increasingly redundant. As Statalist and the SSC and other archives exemplified, users could publish their programs on the Internet as soon as they liked and make fixes and enhancements as soon as they liked. Even with frequent publication rate and short queues, the STB could not possibly match that. Diskettes were phased out to the extent that what was published in the STB could be installed in Stata by users across the net (regardless of whether they were paid subscribers), but this did not affect the major issue. For that and other reasons, the STB was relaunched in 2001 as the *Stata Journal*. Joe Newton maintained continuity as Editor of both and was joined from the launch by Nicholas Cox. The story has been told in greater length elsewhere (Newton and Cox 2003).

The first 64-bit version for Solaris became available in 2001. The second was the 64-bit Linux, and others have followed for Windows and Mac OS X. Another adjustment to bigger sizes was the release of Stata/SE with support for larger datasets.

Stata 8 was the biggest release in the history of Stata so far, and one of the most painful internally, because of the very large mass of code and documentation released all at once. This featured a completely new graphics system. Its principal architect was Vince Wiggins, who oversaw the development team. The new graphics system was the biggest single project so far in the history of the company. Although it came some years later than many users had hoped, the delay was not for lack of much long and hard work. A new graphics system had been developed some years before, but that project was pulled as not good enough. StataCorp struggled for some while to maintain backward compatibility with previous graphics, but reluctantly had to sacrifice that goal. Those who have worked with the new graphics appreciate the flexibility and versatility of the new design, and many users have adopted it, and even built upon it, with enthusiasm.

Another project of almost equal size was the introduction of dialogs in Stata 8. Stata had previously had some dialog functionality, but by the standards of many windowed products, it was rather limited. Naturally, many users, and many more potential users, are accustomed to dialogs in much of their software experience. One of several crucial considerations was the need to make dialogs compatible with the command language; any design that bifurcated Stata into "command language Stata" and "dialog-based Stata" would have violated Stata's design principles. Key to the way that dialogs work is echoing of the equivalent command to the Command window so that beginning users can, it is hoped, see how natural and consistent the language is once you get used to it. Several advanced users reported, at least privately, that playing with dialogs was one of the main ways they first came to grips with the new and much more complicated graphics syntax.

The experience of releasing Stata 8 prompted a change in release policy towards dribbling out more code between major releases. The web-aware character of Stata and the ease with which `updates` are possible for users (except those behind very thick firewalls) encouraged a return to the idea of more frequent releases. Stata 8.1 was sparked by an updated `ml` and Stata 8.2 by some moderately substantial changes to graphics.

16.4   Onwards and upwards

Over the last decade, Stata has fallen into a moderately predictable pattern of a new release every two years. The additions in each release have been so many and so important that even a selection of highlights could not do justice to Stata's progress. Fortunately, Stata itself provides ample documentation of the major details of new releases. You can access this within Stata by first firing up `help whatsnew`. Then you can scroll to the bottom of the Viewer window to access the documentation on new features of each release, both at each new version and in between, over the history of Stata since Stata 6.

Here is a capricious sampling of major additions:

Version 9 included the new matrix programming language Mata; new survey features; linear mixed models; multinomial probit models; and new multivariate analysis commands (with yet more such commands in Stata 10).

Version 10 saw a Graph Editor; logistic and Poisson models with complex, nested error components; exact logistic and exact Poisson regression; limited-information maximum-likelihood and generalized method of moments estimation; more estimators for dynamic panel-data models; systems of nonlinear equations by feasible generalized least squares; and time/date variables in addition to date variables. Especially important for many users was the introduction of Stata/MP, which happened during Stata 9, as the parallel version of Stata for multiple-CPU and multicore computers.

Version 11 featured factor variables as a much improved way of handling categorical predictors, interaction terms, and so forth across many estimation commands; `margins` as a very general postestimation command (made graphical in Stata 12); multiple imputation (also further extended in Stata 12); a new Variables Manager, Data Editor, and Do-file Editor (in Windows, any way); bold and italic text, Greek letters, symbols, superscripts, and subscripts on graphs; and PDF documentation.

Version 12 was marked by automatic memory management; structural equation modeling (further expanded in Stata 13); contrasts and pairwise comparisons; multivariate generalized autoregressive conditionally heteroskedastic, unobserved components model, autoregressive integrated moving average and various filters for time series; business calendars; and easier importing and exporting.

Version 13 broke a longstanding limit of 244 characters to introduce long strings or binary large objects. More statistical innovations focused on treatment effects, new multilevel mixed-effects models, forecasts based on systems of equations, expanded power and sample size calculations, new and extended panel-data estimators, and more for calculating effect sizes. This release was also marked by a new Project Manager, Java plugins, and improved `search` and `help`.

Over recent years the company and its products have grown qualitatively as well as quantitatively. The relaunch of Statalist as a web forum in 2014 (interestingly enough, a development regretted by several users who preferred the longstanding email-based listserver style) was emblematic of efforts to keep pace with newer ways of finding and sharing information. Under the same heading can be mentioned Stata's presence on Facebook, Twitter, and other social media and the very popular launch of Youtube videos. StataCorp has also become increasingly active, and indeed proactive, in off-site training and appearances at disciplinary meetings.

16.5   Pervasive principles

What principles have guided Stata? Let me attempt to distill some of the most pervasive.

Despite modern user interfaces, the heart of Stata remains the command language. Whatever is done via menus or dialogs is, ideally, echoed as a command. The overwhelming emphasis on a command language follows from a firm belief that statistical analysis cannot be reduced to a small series of standard tasks. In particular, smarter statistical users, especially those near the cutting edge of research in many fields, do not want the statistical equivalent of a burger bar with choice from a fixed menu, however appealing the individual dishes may seem to some tastes. They do want to go beyond the menu and go inside the kitchen, not to peel the potatoes or fillet the fish, but to order something not on the menu and discuss it with the chef.

One key to good Stata practice is combining do-files and log files to produce reproducible research. A do-file documents a series of analytical steps. A log file documents a series of results. Any statistical project can go through many iterations, digressions, and dead ends, but one aim should remain the production of the do-file and the log file, documenting what you would have done at the outset had you been smart enough and well-informed enough to know what you now know. It hardly needs underlining that this combination depends on a command language.

The executable provides a kernel for the statistical operating system that is Stata. It gives the functionality that users would usually find too difficult or time-consuming to write themselves. The executable in 2015 is much, much bigger than the first executable compiled in 1984; no surprise there. But a more important measure is the proportion of Stata written in Stata or Mata as ado-files or other program files. Asymptotically, it will approach 1!

Maintaining platform independence is important for Stata. Why does StataCorp do this when almost all users use just one of the platforms supported (and most of those use one particular platform)? The answer lies in a mix of reasons that are technical, business, and aesthetic. It seems neither prudent nor tasteful to yoke Stata to any particular platform.

Maintaining backwards compatibility as Stata is extended is also vital. StataCorp's marketing style reflects a hope that users will stay loyal over the long run, evidenced by the high fraction who upgrade on each new release. To encourage and reward that, StataCorp tries to minimize changes that will break users' programs, do-files, or habits. This is not always easy, and mechanisms have been developed to allow this, most notably version control.

Statistical software differs from most other software in one key respect: the way that users need, and indeed contribute to, an extensible language. Such software, however good, still leaves many users with statistical needs they want to satisfy, but many of them do have the ability to write programs to meet those needs. Stata users are often aware of methods, even whole fields, not yet covered by Stata, as is inevitable given the size and complexity of statistics and the large number of sciences using statistics. Stata has gained considerably from what users themselves have done. For this to be true, it is not essential that all Stata users become Stata programmers. Just a few hundred who have put their work into the public domain have been sufficient to give other users extra leverage and to have impacted Stata itself. A user community contributing to Stata directly and indirectly drives a mighty feedback loop, propelling Stata's continued growth.

16.6   Acknowledgments

Bill Gould (*sine quo non*) provided most of the raw material for this essay. He and Pat Branton also provided valuable access to copies of early documentation.

References

Anagnoson, J. T. 2005. The history of StataQuest. *Stata Journal* 5: 41–42.

Cox, N. J. 2005. A brief history of Stata on its 20th anniversary. *Stata Journal* 5: 2–18.

Gould, W., J. Pitblado, and B. Poi. 2010. *Maximum Likelihood Estimation with Stata.* 4th ed. College Station, TX: Stata Press.

Gould, W., and W. Sribney. 1999. *Maximum Likelihood Estimation with Stata*. College
Station, TX: Stata Press.

Hamilton, L. 2005. A short history of Statistics with Stata. *Stata Journal* 5: 35–37.

Hilbe, J. M. 2005. The birth of the bulletin. *Stata Journal* 5: 39–40.

Newton, H. J. 2005. A conversation with William Gould. *Stata Journal* 5: 19–31.

Newton, H. J., and N. J. Cox. 2003. The Stata Journal so far: Editor's report. *Stata
Journal* 3: 105–108.

Royston, P. 2005. Stata at 20: a personal view. *Stata Journal* 5: 43–45.

Timberlake, T., and N. J. Cox. 2010. Ana Isabel Palma Carlos Timberlake (1943–2009).
Stata Journal 2010: 9–10.

16.7   About the author

Nicholas Cox is a statistically minded geographer at Durham University. He contributes
talks, postings, FAQs, and programs to the Stata user community. He has also coauthored 15 commands in official Stata. He was an author of several inserts in the STB and
is an editor of the *Stata Journal*. His Speaking Stata columns on graphics from 2004 to
2013 have been collected as *Speaking Stata Graphics* (College Station, TX: Stata Press,
2014).